Ethics in Artificial Intelligence and Information Technologies

Gabriela Arriagada-Bruneau
Institute of Applied Ethics
Institute of Mathematical and Computational Engineering
Pontificia Universidad Católica de Chile, Chile

Claudia López
Department of Informatics
Universidad Técnica Federico Santa María, Chile

Marcelo Mendoza
Department of Computer Science, Faculty of Engineering
Pontificia Universidad Católica de Chile, Chile

CRC Press
Taylor & Francis Group
Boca Raton London New York

CRC Press is an imprint of the
Taylor & Francis Group, an **informa** business
A SCIENCE PUBLISHERS BOOK

First edition published 2026
by CRC Press
2385 NW Executive Center Drive, Suite 320, Boca Raton FL 33431

and by CRC Press
4 Park Square, Milton Park, Abingdon, Oxon, OX14 4RN

CRC Press is an imprint of Taylor & Francis Group, LLC

Library of Congress Cataloging-in-Publication Data (applied for)

ISBN: 978-1-032-59418-7 (hbk)
ISBN: 978-1-032-59422-4 (pbk)
ISBN: 978-1-003-45462-5 (ebk)

DOI: 10.1201/9781003454625

Typeset in Times New Roman
by Prime Publishing Services

Preface

The development of new information technologies is significantly boosted by the adoption of Artificial Intelligence (AI) based technologies. The possibilities offered by AI are unsuspected and force us to reconsider various aspects of our lives, including content creation, social media interaction, and educational methods in schools and universities. Consequently, it is inevitable that the influence of AI-based breakthrough technologies will grow in the coming decades.

In this book, we address the challenges posed by the adoption and development ofthese technologies and their impact on people. In this context, the ethical considerations, scope, and impact of technology on people are crucial. To start with, the book discusses the ethical aspects of AI, presents a socio-technical approach to integrating Ethics into AI projects, and outlines perspectives around feminism, sustainability, and labor transformation. Next, the concepts of fairness, accountability, and transparency are introduced, exploring their implications for developing information systems like recommender systems, with a focus on data privacy aspects. Then the book focuses on the relevance of atural language processing systems, highlighting debias strategies and evaluation methodologies. The scopes of fairness-based approaches for ChatGPT and other generative language models are also introduced. Finally, advanced topics that include the relationship between AI and disinformation are addressed, including a discussion on the scope of news-generative models that produce deep fakes.

The book ends with a discussion of the perspectives and challenges in the area. This book is devoted to an audience of advanced undergraduate and graduate students from all disciplines related to information systems. It is also helpful for researchers and practitioners interested in the subject.

Contents

List of Figures

List of Tables

ETHICS AND AI

I

Chapter 1

What is AI Ethics?

1.1 Introduction

In this introductory chapter, we highlight key aspects of applied ethics in the context of Artificial Intelligence (AI). We explain what AI ethics entails, its objectives and various ways researchers can contribute to advancing this field. More specifically, we wish to show our perspective as an interdisciplinary group that interacts with AI ethics from different standpoints. In part, this is a response to a gap identified by Paula Boddington in the introduction of her book "AI Ethics: A Textbook," where she claims there is "a pressing need for contrasting voices to contribute to this field and to recognize the complexity of AI ethics" [34, p.1]. Acknowledging this need, we wish to offer a view to understand and study AI Ethics that comes from our own needs, experiences, and limitations as researchers. Hence, in this book, we will combine our expertise in applied ethics, Human-Computer Interaction (HCI), and Natural Language Processing (NLP) to present different dimensions of ethical issues in AI.

First, we will explain why AI Ethics has become increasingly relevant and why it is a challenging endeavour. Then, we will explain how we understand AI ethics, the position of critical voices around AI ethics, and the line of research that we follow to seek approaches to address ethical challenges that emerge in AI projects. To conclude, we will highlight the key ideas shaping the most recent debates on AI ethics, setting the stage for the discussions in the subsequent chapters.

1.2 Why has AI ethics become so relevant, and why does it remain challenging?

Over the past decade, AI ethics has evolved from an emerging field in Applied Ethics to a widespread necessity for researchers, companies, governments, and developers. Society calls for ethical standards to guide the rapid and expansive technological revolution driven by AI technologies. This parallels how bioethics became a focal point in Applied Ethics when dilemmas like cloning, euthanasia, abortion, and genetic testing arose. These dilemmas required experts who were not only knowledgeable in ethical theories but capable of applying them to medical practice, thereby establishing bioethics as an inherently interdisciplinary sub-field.

AI ethics surface to address the challenges of developing and deploying AI technologies in society, including various application scenarios. To illustrate its increasing relevance, consider how the Google count of academic papers with "AI" and "Ethics" in their titles has grown since 1985 (see Figure 1.1). Borenstein et al. [37] conducted a search to demonstrate that while exploring ethical issues in AI may now seem commonplace, it was not always so. In their original figure, they showed the number of articles that appeared in Google Scholar under the tags ("ethics" or "ethical") and ("AI" or "artificial intelligence"). We performed an updated search with the same criteria up until 2023 (Figure 1.1) and observed that the increasing trend continues. The consolidation of the relevance of AI ethics seems ubiquitous nowadays—at least in the academic world.

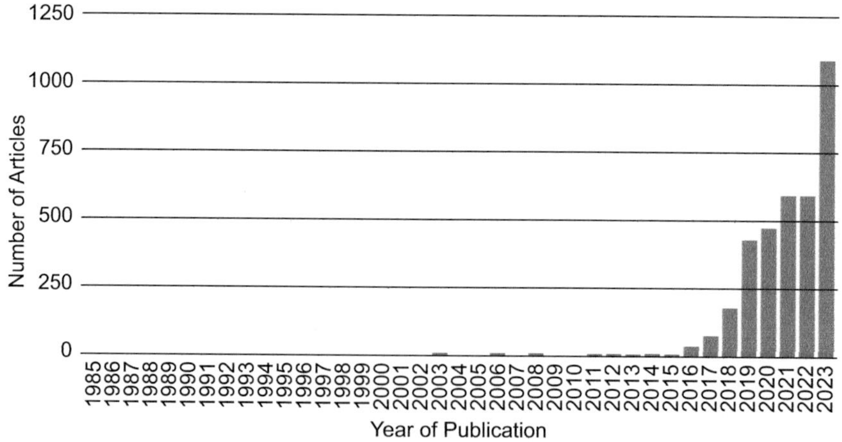

Figure 1.1: Update (until 2023) of the Google Scholar search conducted by Borenstein et al. [37].

What is interesting about this historical look in the scholarly discourse is the origin of this "hype" for AI Ethics, which was fed by famous cases that occurred in 2016, such as the "racist algorithm" of COMPAS (Correctional Offender Management Profiling for Alternative Sanctions). Real-life cases significantly influenced the integration of ethics into AI development. AI transitioned from a niche area to a mainstream technology. With its evolution and the advent of big data and advanced computational capabilities the societal impact of AI made ethical considerations in its development and implementation into an unavoidable ethical discussion.

This discussion quickly became challenging because concepts with a long tradition in moral philosophy, like fairness, have taken on new meanings and contexts in AI applications. Let's explore a bit deeper into the COMPAS case to illustrate this point.

This risk assessment algorithm developed by Equivant (formerly Northpointe, Inc.) is designed to predict an individual's likelihood of recidivism. This prediction's purpose was to assist judges and inform parole officers in deciding if defendants awaiting trials are too risky and, therefore, not assignable a release on bail — i.e., if their predicted recidivism score is too high, they should be deemed risky. The algorithm's training data comes from various sources. A portion of the training data is derived from historical risk assessments made by judges in the United States. The data also includes responses from questionnaires filled out by defendants and evaluations by correctional officers regarding their perceived risk of the offenders. The trained AI model assigns defendants a score ranging from 1 to 10 with higher scores indicating a greater risk of recividism.

The ethical debate was sparked by a group of investigative journalists from ProPublica, who reported that black offenders were more likely to be labelled as higher risk than white offenders, indicating a bias in the algorithm against black individuals [9, 179]. ProPublica analyzed risk scores assigned from 2013 to 2014 and compared them with individuals actual reoffenses within the next two years. For example, a black defendant, B.P. was labelled high risk by COMPAS (score of 10/10) despite having only one prior offence of resisting arrest without violence and he committed no subsequent offences within two years. Conversely, a white defendant, V.P., was labelled low risk (score of 3/10) despite having a criminal history of two armed robberies and one attempted armed robbery and he committed one grand theft within two years. These cases were not isolated examples.

An analysis of false positives and false negatives across a sample of more than 6,000 people showed that while the algorithm was able to predict recidivism correctly 61% of the time, it made errors differently across races. See the confusion matrices for Black and White defendants in Tables 1.1 and the error rates by race in Table 1.2. Specifically, the rate of false positive misclassifications—where individuals were predicted to have a higher risk of recidivism, but they did not reoffend in the next two years—was nearly twice as high for Black defendants compared to White defendants (44.8% vs 23.5%).

Conversely, the rate of false negative misclassifications —where people were classified at a lower risk of recidivism, but they did, in fact, reoffend— was considerably higher for White than Black reoffenders (47.7% vs 28%). From these findings, one can deduce that the algorithm overestimated the recidivism risk for Black defendants and underestimated it for White ones, leading to the conclusion that the algorithm was biased against Black defendants [9].

Table 1.1: Confusion matrices for Black and White defendants

	Black defendants			**White defendants**		
	Predicted risk			**Predicted risk**		
	Low	**High**	**Total**	**Low**	**High**	**Total**
Did not reoffend	990	805	1795	1139	349	1488
Re-offended	532	1369	1901	461	505	966
Total	1252	2174	3696	1600	854	2454

In response to ProPublica's claims that the algorithm was 'being racist' and unfair, Equivant argued that ProPublica's use of model error metrics such as false positive and false negative rates to evaluate racial bias was misguided. Equivant contended that other measures, such as positive and negative predictive values (and their complements), should be used instead. Their main argument was that false positive and false negative rates are influenced by differences in the base rates of the behavior being studied. They pointed out that the distinct rates of recidivism of Black and White defendants (0.51 and 0.39, respectively) impact the error metrics, resulting in a higher false positive rate among Black defendants due to their higher base rate.

Equivant asserted that positive and negative predictive values, which assess the algorithm's accuracy in making positive or negative predictions across different races, provide a more accurate measure of the algorithm's racial bias. This shifts the focus from errors related to actual behavior to errors associated with specific predictions (refer to calculations in Tables 1.2 and 1.3). Equivant suggested that for an algorithm to be fair it should demonstrate predictive parity. This means that the algorithm should have an equal ability to identify recidivists and non-recidivists among both Black and White populations. Using the positive predictive values (37% for Black defendants, 40.9% for White defendants), Equivant concluded that the likelihood of recidivism among high risk defendants is similar across races. This, they argued, demonstrates that their algorithm achieved predictive parity. Therefore, according to Equivant, ProPublica's claims were unfounded because the pattern of false positives (percentage of non-recidivists misclassified as recidivists) affecting blacks over whites "does not show evidence of bias, but rather is a natural consequence of using unbiased scoring rules for groups that happen to have different distributions of scores" [80, p.8].

Table 1.2: ProPublica's numbers to support its claim about racial bias

	Black defendants		White defendants	
False positive rate	805/1795 * 100	(44.8%)	349/1488 * 100	(23.5%)
False negative rate	532/1901 * 100	(28%)	461/966 * 100	(47.7%)

Table 1.3: Equivant's numbers to support its claim about predictive parity

	Black defendants		White defendants	
1 - PPV	(1-1369/2174) * 100	(37%)	(1-505/854) * 100	(40.9%)
1 - NPV	(1-990/1522) * 100	(35%)	(1-1139/1600) * 100	(28.8%)

PPV: Positive predicted value
NPV: Negative predicted value

A legal case, 'State of Wisconsin v. Loomis' 881 N.W.2d 749 [221] in the United States, added a legal precedent to the COMPAS case. The defendant claimed there was a violation of his due process rights by using this risk assessment algorithm. Loomis and his attorneys requested a subpoena because they were unable to obtain specific details about how the algorithm generated his individual score. According to the court, COMPAS was not required to present details about its algorithms, as they were classified as trade secrets. When Loomis appealed the use of COMPAS, the court claimed that it was accurate to say that Loomis' sentence would have been exactly the same regardless of the use of COMPAS. The court reiterated that the use of COMPAS helped corroborate the sentence and, therefore, there was no interference in the decision-making. In this legal case, the Supreme Court dismissed Loomis' claim, citing the availability of the COMPAS reports to both the State and the defendant (with results provided before the trial) and the opportunity for the defendant to revise the questionnaire responses. The court's evaluation of the algorithm focused on the accuracy of the defendant's responses to the questionnaire (a portion of the algorithm's input data) and the opportunity to verify them. Additionally, it noted that the final sentence, would have remained unchanged regardless of the COMPAS predictions.

Researchers have contended that assessing an algorithm's fairness solely based on input data accuracy is inadequate. This approach overlooks other crucial aspects of data processing and analysis [310]. Washington [310] argues that focusing solely on a single aspect of data quality overlooks the broader complexities involved in evaluating algorithms, especially within the public sector. For example, concerns about the visibility of decisions and the difficulties of scrutinizing procedures are shielded in the opacity surrounding algorithms. Hence, at first glance, what appears to be an issue about fairness or justice is also related to transparency.

Another crucial concern regarding COMPAS is related to the types of questions included in the risk assessment questionnaire, which serve as input for the algorithm. Some of these questions raise potential issues with fair representation and stigmatization, for example:

- Based on the screener's observation, is this person a suspected or admitted gang member?
- How many of your friends/acquaintances have ever been arrested?
- Were you ever suspended or expelled from school?

Considering these questions as relevant input might inherently reflect and perpetuate biases, calling into question, from another perspective, the fairness of the algorithm's application.

Hence, we have a challenging scenario in which to establish a criterion for fairness. On the one hand, Equivant justified the outcome as not racist based on predictive parity and technical analysis, claiming that it is equally fair at predicting across races. On the other hand, ProPublica's argument is centred on the outcome, showing COMPAS' ethical blind spot: it disproportionately affected black offenders by overestimating their risk of reoffending, even though race was not included explicitly as a variable in the predictions. In turn, the Supreme Court decided to limit the criteria to accurate measurements of the input data, while some input data can be questioned due to its legitimacy.

The COMPAS highlights the complexity of debating fairness in AI considering various dimensions such as technical, ethical, and legal aspects. It demonstrates the challenge of understanding fairness through a single lens because it intertwines with many aspects, such as accuracy, bias, procedural aspects, discrimination and transparency. COMPAS illustrates how discussions about algorithmic fairness can quickly evolve into a never-ending debate on the meaning of fairness.

The challenge of discussing fairness in AI also involves the difficulties of operationalizing and implementing technical notions of fairness to algorithmic models. The ethical input often mirrors structural inequalities in society, the criteria of existing legislation can either dismiss or prioritize specific definitions to evaluate fairness, and the required criteria to evaluate the success of AI development can influence differently the measures of fairness. Moreover, these various dimensions of fairness are often designed and limited to their disciplinary domains. While they may theoretically influence other dimensions they are seldom conceptualized as an integrated whole given the practical limitations involved.

This case analysis highlights not only the complex task of addressing AI-related ethical questions but also the growing significance AI ethics has acquired in recent years. This is especially true for topics that directly impact people. AI applications are being used to enhance agricultural efficiency, study wildfires and earthquakes, and monitor ocean tides. They also extend to domains that have a more direct link to human needs and well-being, such as treating dementia patients, managing and surveilling employees, tracking students' academic progress, monitoring infected individuals during the COVID-19 pandemic, for navigation maps, assistant chatbots, and criminal justice systems, like the COMPAS case.

There is no doubt that AI ethics is now surfing through a hype wave; however, there are fundamental ethical questions rooted in concerns from longstanding ethical debates, such as issues about fairness, transparency and responsibility. Therefore, one of the core challenges is translating these ethical debates into meaningful practices. Given that AI technologies have become integral to numerous aspects of daily life—from personal interactions with virtual assistants like Siri and Alexa to more complex systems such as autonomous vehicles and predictive policing—this widespread adoption necessitates a thorough examination of the ethical implications of AI technologies on society. It also requires a careful consideration of how professionals developing AI integrate and understand AI ethics.

As technology evolves and becomes an integral part of society, influencing decision-making, scientific discovery, and problem-solving awareness of the importance of ethics in AI development continues to grow. Despite the rapid increase in AI ethics research, translating ethical principles into meaningful operational practices remains a significant challenge. For example, numerous research articles, reports, books, and guidelines define principles for ethical AI, such as fairness. How they define fairness might share certain commonalities, such as avoiding bias or preventing discrimination [157, 101]. However, these principles advocating for "fairness" often lack specific guidance on addressing fairness-related issues. In the previously discussed COMPAS case, a principle advocating for "avoiding bias" cannot be translated into specific practices when the three domains of fairness—technical, ethical, and legal—are being discussed, entangled, and interact with each other. Hence, part of the challenge that we currently face about AI ethics is no longer rooted in proving its relevance but rather breaching the practical, methodological, and professional gaps for AI ethics to be tangibly valuable, i.e., offering concrete approaches, frameworks, and measures that are a cohesive integral part of AI development and implementation.

But before continuing our discussion, it is essential to understand what AI ethics entails to convincingly argue why it should be studied.

1.3 What will we understand as AI Ethics?

To define AI ethics, we first need to know what ethics is. Moral philosophy is the area that studies ethics, which is concerned with moral phenomena, i.e., normative questions about what people should do, based on reasons and judgements that allow one to argue when an action can be morally right or wrong. Within the systematic study of these moral phenomena, there are three main areas of study: metaethics, normative ethics, and applied ethics.

Metaethics concerns fundamental questions about ethical matters, in other words, the nature of morality. It offers definitions to grounding concepts such as goodness or badness, the objectivity of morality, or the meaning and origin of moral judgements. Thus, metaethics studies how we think, know, and conceptualize morality, informing normative ethics.

Normative ethics explores universal principles to guide actions from various theoretical perspectives such as consequentialism, virtue ethics, deontology, and the ethics of care. It examines the morality of actions, motivations, and character traits addressing questions like "What makes murder morally wrong?" or "Is stealing ever morally acceptable?" This field aims to identify general principles that justify moral actions.

Applied ethics addresses practical moral questions related to specific issues like animal rights, climate change, and AI development. Unlike normative ethics which focuses on universal principles, applied ethics considers practical implications of morality and its societal impact. This branch of philosophy extends the discussion to particular situations and contexts, including the realm of AI ethics.

AI ethics encompasses several aspects of AI development. For example, discussions often focus on roboethics, which deals with AI software integrated into physical machines that interact with their environment, primarily through sensors, leading to discussions about robot rights [60, 128]. Similarly, there is a dedicated field for human-robot interaction, examining our social and psychological relationships with robots [112, 87] as well as specific application debates for social robots [72], such as care robots [58] [105] and sex robots [73, 176] or other applications in the military [109, 138] and healthcare sectors [280, 314]. Closely related are discussions about machine ethics, which focus on moral machines [178, 306] and artificial moral agents [50, 125, 210].

Another area of debate in AI ethics revolves around specific advancements in the AI field and their societal implications. For example, ethical research on autonomous vehicles often references the longstanding ethical dilemma known as the "Trolley problem" [290, 291]. This dilemma involves deciding whether to pull a lever to divert a trolley from a track where it would kill five people to a different track where it would kill only one person. These hypotheticals are often used as a type of thought experiment to contest what would be the right choice. The dilemma aims to explore our intuitions concerning acting (killing one person) or letting something happen (letting five get killed). These hypothetical scenarios have been included in discussions about the ethical design of autonomous vehicles, considering various factors and prioritizing cultural differences when programming and training these vehicles. Thus, when these autonomous vehicles crash, for example, part of the research examines how to make them "crash morally" [309].

Other prominent areas of focus include AI and healthcare, where discussions revolve around classification algorithms for diagnosing cancer patients [145], the use of AI for mental health diagnosis [165] and the ethical issues related to employing digital biomarkers and AI to detect early dementia [104]. AI is also being applied in education, such as through AI-driven mobile apps [173] AI's general pedagogical impacts are also being studied [130]. Additionally, automation poses risks to employment, leading to employment polarization [55, 293]. This is characterized by slow wage growth for lower-skill workers, who are increasingly replaced by AI driven productivity, while highly educated workers see increased wages.

Thus, here we will understand AI Ethics as a sub-field of Applied Ethics concerned with establishing good practices for the ethical development and implementation of AI. As a multi- and inter-disciplinary field, AI Ethics draws insights from various disciplines, including but not limited to ethics, sociology, computer science, engineering, philosophy, psychology, medicine, and law. Studying AI ethics necessitates collaboration among professionals from various fields to develop guidance and methodologies that foster a harmonious ethical relationship between AI and society.

1.4 Are there critical voices regarding AI Ethics?

Like all disciplines, AI ethics faces challenges and limitations. One of the core criticisms is the effectiveness of defining AI principles, a strategy commonly adopted by organizations, countries and multilateral agencies worldwide. Several authors have criticized the "uselessness" of AI ethical principles and guidelines. Mittelstadt [209], for example, argues that rather than offering concrete, targeted recommendations, "many initiatives, particularly those sponsored by industry, have been characterized as mere virtue signalling intended to delay regulation and pre-emptively focus debate on abstract problems and technical solutions" [209, p.501]. This perspective poses that ethical standards are abstract, ignoring the normative and political challenges of key AI concepts such as privacy and fairness.

Hagendorff argues something similar, claiming that "AI ethics—or ethics in general—lacks mechanisms to reinforce its own normative claims" [129, p.99]. Hagendorff stresses that based on the implementation of ethical principles, institutions can take an easy way out and establish their own ethical guidelines, promoting an illusion of self-regulation. Thus, employing AI ethics principles as a facade for ethical compliance, known as ethics washing, is a concern. Bietti describes this phenomenon in the AI context as the situation where 'ethics' is increasingly associated with technology companies' self-regulatory efforts and superficial displays of ethical behavior." [31, p.210]

More recently, Munn criticizes AI ethics for providing "meaningless principles, isolated principles, and toothless principles a gap between principles and practice" [213, pp.869–870]. He emphasizes that AI ethical principles offer ambiguous guidance, which allows existing practices to continue maintaining the industry's status quo. Munn also highlights the reluctance of engineers to engage with ethical questions, which is symptomatic of a larger, more pervasive problem within the tech industry, which is making "unethical AI the logical byproduct of an unethical industry" [213, p.871].

Similarly, Griffins et al. [108] interviewed 40 AI developers and found that almost half of them reported that being a developer was "neither ethical nor unethical" [108, p.4]. This neutral appreciation of their profession sees AI development not as an ethical endeavour but rather as an action performed to get paid and tasks concerned with efficiency, optimization, and more technical decisions. A key issue here is that this perception of AI as ethically neutral can often stem from a lack of knowledge and awareness of what can be identified as an ethical aspect of AI development.

Furthermore, our own work has found evidence that AI developers and researchers tend to think that ethics tell us what not to do [186]. Many of the researchers and developers highlighted that they perceived ethics as negative, as a restrictive imposition to research and innovation, turning ethics into an obstacle more than an ally. This view is commonly brought up by people when their main relationship with ethics has been through ethical committees or principled AI, prompting a limited understanding of ethics. Despite recognizing the relevance of AI ethics, their main concerns were based on the lack of guidelines and concrete mechanisms for putting ethics into practice.

Hence, these criticisms highlight the importance of developing new approaches that are more practical and contextual and that provide a more robust structure for principles to be effective. This is challenging, not only because AI ethics is a new field but also because of the interdisciplinary collaborations needed to come up with alternatives, as well as the evolving development landscape of AI. Thus, more than a limitation per se, these criticisms against AI ethics principles demand more comprehensive and holistic ways to understand the challenges that it brings.

Acknowledging these criticisms, it becomes essential to discuss why this area of study matters and why it is relevant for different professionals and scholars to get acquainted with it. Our answer is based on the fact that ethics can be understood not only as an individual decision-making process but instead as a "need to justify or explain ourselves to others. Ethics is the study of what actions really can be defended under scrutiny." [29] In an interdisciplinary field like AI, communicating and appropriately justifying decisions about AI development and implementation is increasingly crucial as AI is increasingly being adopted into different domains in society.

1.5 How can we integrate Ethics into AI?

In response to these criticisms of AI ethics, researchers have advocated for embedding ethics into AI. This can be achieved by ensuring that AI systems embody certain values [298], incorporating ethicists into AI development teams [194], or integrating ethics into machine learning courses [260].

Other researchers, like Johnson and Verdicchio [159, 158], offer a critique against the simplistic notion of "embedding" ethics into AI, problematising its meaning. They claim that merely adding ethics to technology, that is, "Ethics + AI = Ethical AI," is a perspective that assumes that ethical principles can straightforwardly be encoded into AI, transforming it into ethical AI, thus committing an additive fallacy. The authors' argument is that for the addition of AI and ethics to be possible, these two areas must share ontological characteristics—that is, they must be of compatible natures. But, since AI has a computational basis, this would imply that ethics, too, must be rendered computational to be integrated with AI. This would mean that ethical principles can be truly captured in computational forms, which does not seem to be plausible. This discrepancy raises doubts about the feasibility of directly translating ethical principles into computational algorithms.

Therefore, they propose that a broader understanding of AI as part of sociotechnical systems can avoid this additive fallacy. This understanding involves considering that AI is not just a set of computational tools but operates within complex networks of human relationships, societal norms, and organizational practices. Ethical considerations in AI thus transcend mere computation and involve the broader sociotechnical context in which AI systems exist. This perspective shifts the focus from trying to make AI intrinsically ethical to considering how AI practices influence and are influenced by societal values and norms.

In this regard, ethics is not telling us what not to do, but instead, it is helping us establish a viewpoint from which AI is considered a sociotechnical tool that can be better developed and implemented when ethical considerations are inherent to the demands of building and using an AI system. Thus, creating AI involves more than just technical solutions; it demands engagement with the ethical dimensions of the domains and industries where AI is applied. AI experts are called to not only develop technology but also critically engage with the broader ethical and social implications of their work— instead of embedding ethics into the computational ontology of AI.

Thus, AI ethics is crucial not only for ensuring that AI systems are developed and implemented responsibly but also for fostering a culture where ethical considerations are seen as fundamental and inherent to the technological innovation process. This culture change challenges developers, policymakers, and stakeholders to critically engage with the ethical implications of their work, encouraging equitable and socially beneficial AI. The importance of AI Ethics lies in its ability to bridge the gap between technological capabilities and societal values, ensuring that AI serves as a tool for positive societal transformation rather than a source of contention and division. For this, diverse interdisciplinary sociotechnical methodologies (see Chapter 2, section 2.2) can be used to engage with creating ethical AI.

1.6 What are the main concerns of AI Ethics?

To frame our conversation about the main concerns in AI Ethics, we will utilize the commonly used, albeit critiqued, method of structuring discussions and practices in AI ethics around principles. We chose this strategy primarily to connect with previous research, allowing us to revisit essential concepts and identify both distinctions and relationships between them.

Various organizations have established principles intended to guide the development of AI. This principles-based approach has been adopted by professional groups like IEEE [272], corporations such as Google [239], IBM [146], Telefonica [288], and Microsoft [204] as well as governments and multi-national institutions, including UNESCO [295], the OECD [65] and the United Nations [142]. By early 2024, 42 countries have committed to the OECD AI principles, and the UN AI resolution has been unanimously adopted by all 193 member nations.

Several scholars have compiled summaries and analyzes of these principles, as detailed in Table 1.4. Many of these review articles identify recurring themes, such as privacy, transparency, accountability, and fairness. However, certain reviews highlight specific elements more prominently than others, influenced by the sources they examine or the sectors they observe. These reports consider perspectives from academia, government, or industry, offering a variety of angles and aspects for consideration.

Drawing from these surveys, we propose a categorization for analyzing AI principles (see Figure 1.2). This categorization is the outcome of a literature review, a qualitative diagnostic study, and a participatory process conducted at the National AI Center, CENIA, in Chile [186]. We divide principles into two main categories. At the top, we put three core principles that cover broader ethical implications in AI: sustainability, human rights, and human control. These three principles comprise impacts, risks, and concerns related to fundamental aspects of human development based on respect for human flourishing, autonomy, and dignity. Therefore, this first set represents a foundation for principled-based discussions. Here, we include:

- **Sustainability**, which involves ensuring vital living conditions and protecting the environment for future generations [102]. This principle addresses the environmental impact of the AI lifecycle [157], which

Table 1.4: Review and principles summary of AI Ethics research

Review	Principles
Fjeld, J., Achten, N., Hilligoss, H., Nagy, A., and Srikumar, M. (2020). Principled AI: A Map of Ethical and Rights-Based Approaches to Principles for AI.	Privacy, accountability, fairness, security, professional responsibility, promotion of human values, transparency, human control of technology.
Khan, A. A., Badshah, S., Liang, P., Khan, B., Waseem, M., Niazi, M., and Akbar, M. A. (2021). Ethics of AI: A Systematic Literature Review of Principles and Challenges.	Transparency, privacy, accountability, equity, autonomy, explainability, fairness, non-maleficence, human dignity, beneficence, responsibility, safety, data security, sustainability, freedom, solidarity, prosperity, effectiveness, accuracy, predictability, interpretability.
Jobin, A., Ienca, M., and Vayena, E. (2019). The Global Landscape of AI Ethics Guidelines	Transparency, fairness and justice, doing no harm, accountability, privacy, doing good, freedom and autonomy, trust, sustainability, dignity and solidarity.
Zeng, Y., Lu, E., and Huangfu, C. (2018). Linking Artificial Intelligence Principles.	Humanity, collaboration, sharing (equity), justice, transparency, privacy, security, protection, accountability, AGI/ASI.
Floridi, L., Cowls, J., Beltrametti, M., Chatila, R., Chazerand, P., Dignum, V., Luetge, C., Madelin, R., Pagallo, U., Rossi, F., Schafer, B., Valcke, P., and Vayena, E. (2018). AI4People—An Ethical Framework for a Good AI Society: Opportunities, Risks, Principles, and Recommendations.	Beneficence, non-maleficence, autonomy, justice, explainability.
Smit, K., Zoet, M., and Van Merten, J. (2018). A Review of AI Principles in Practice.	Human enhancement, beneficence, reliable, human-centered. Autonomy, equality (design and execution), traceability, human dignity, human rights, transparency, democratization, privacy, security, security (design and execution), collaboration, responsibility, comprehensibility, responsible data use, accuracy, and education and promotion.

encompasses the significant use of energy and water in AI development and application [27, 188], as well as the extensive extraction of natural resources for AI infrastructure [67]. Ethical AI initiatives should assess their energy consumption and overall impact on the environment, striving to reduce their carbon footprint, optimize energy efficiency, and utilize renewable energy sources for their infrastructure. Certain authors also consider the notion of sustainable communities to be a fundamental aspect of the sustainability impact of AI [181], which involves evaluating and continuously reviewing the anticipated and actual effects

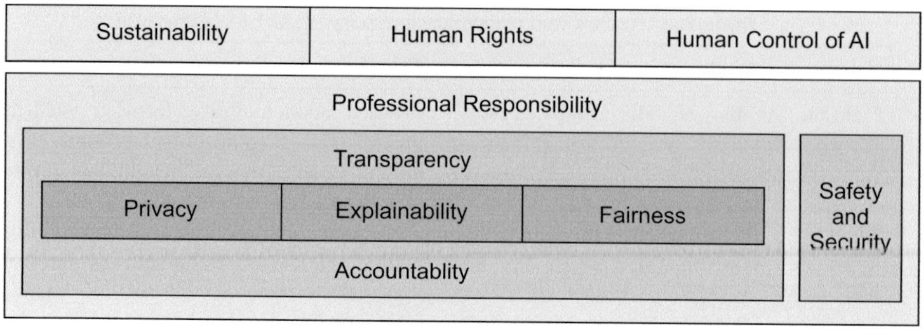

Figure 1.2: Categorization of AI principles [186].

on the communities impacted by AI, including the intended and unintended consequences.

- **Human Rights** have been utilized as a framework to capture the broad spectrum of positive and negative impacts that AI could have in people's lives. Protecting human rights has become essential in response to observed disparities in performance metrics such as error rates among different gender or racial groups, particularly in fields affecting healthcare access [222], legal processes [9], and recruitment decisions [174]. The first UN resolution on AI emphasizes the importance of upholding human rights to ensure that AI systems are secure, safe, and trustworthy [15]. Human rights encompass the rights to equality, dignity, freedom of movement, property ownership, and access to healthcare and employment, among others. Scholars have highlighted the pivotal role of human rights as a universal and normative force, pointing out their significance in guiding practices even where specific national AI legislations are lacking [7]. Another related approach emphasizes human values and flourishing, underscoring the role of AI in supporting the progress of human civilization [101]. This also connects to community-oriented concepts tied to the principle of sustainability.
- **Human control of AI** emphasizes the need for human supervision of AI systems as an essential measure to maintain human autonomy [98]. This involves preventing AI from autonomously making critical decisions, guaranteeing the capacity to intervene in AI operations as needed, and ensuring that decision-making authority remains with humans [101]. Although this principle may seem especially applicable to fully autonomous AI systems, its relevance extends to situations where humans are involved in processes mediated by AI. In such contexts, it is critical to determine whether individuals, particularly those affected by AI, can effectively intervene. Such intervention could range from reviewing and providing feedback to the system to challenging AI decisions and opting out of an AI system. Moreover, it is essential to

examine whether decision-makers who utilize AI-generated information, such as judges who use recidivism risk scores, can critically assess these outcomes, acknowledging their limitations and considering the context. Research has shown that people tend to overly trust AI-generated results [152]. Even before the rise of machine learning, researchers had documented how automation over-influences people's decision-making [275], which complicates the effective implementation of this principle and its implications for accountability [274]. Moreover, this principle aims to maintain autonomy by ensuring that individuals can make their own decisions without being unduly influenced by AI, such as through AI-driven persuasion or nudging. It also advocates for requiring user consent prior to interacting with an AI system and for involving users and other stakeholders in the development and feedback stages of the AI lifecycle.

Subsequently, our categorization outlines various AI principles under the overarching theme of professional responsibility. This emphasizes the importance of the individuals behind AI technologies and their responsibility in decision-making processes that affect both AI implications for sustainability, human rights and human control of AI and how other principles are implemented throughout the AI lifecycle. Within the context of professional responsibility, we make distinctions between safety and security concerns and other principles that are more notably affected by the lack of methodological and model transparency in AI development. For this reason, we feature transparency as an overarching principle that encompasses and interlinks with all the principles we place within it, such as accountability, privacy, explainability, and fairness. Detailed descriptions of each of these principles follow:

- **Professional responsibility** refers to the critical role of individuals involved in the decision-making, design, development and deployment of AI. It emphasizes the expectation for these professionals to address a wide range of issues, from ensuring systems performance considering the long-term effects of their decisions [101] on core principles such as sustainability and human rights. Professional responsibility also includes methodological aspects, such as conducting responsible design practices, which have emerged as an approach that considers ethical aspects from the beginning of the lifecycle, not as an afterthought or corrective measure. Privacy by design [265] is one example of these approaches. This principle also calls for guidelines of conduct for individuals and configurations of AI teams to more effectively navigate the ethical complexities of AI. This entails promoting scientific integrity and fostering collaborations among various stakeholders throughout the

AI lifecycle, including those affected by AI, to harness diverse expertise for framing problems, devising solutions and identifying potential risks.

- **Transparency** is one of the most mentioned AI principles. A prevalent method in AI today is machine learning, which generates models by "learning" from a vast collection of data or examples. For instance, in the case of COMPAS, a machine learning algorithm was used to learn a model that predicts a recidivism risk score based on historical data from individuals who have gone through the judicial process. However, this "learned" model is likely to be intelligible for humans, a common characteristic of models developed through machine learning, especially those generated through deep neural networks (or deep learning). This is why the AI models are often called opaque or black boxes. This opacity indicates that humans, including the developers themselves, may not fully grasp how the model arrives at its outcomes or predictions. Humans do know, however, that the model was learned through an algorithm that seeks to optimize a function based on the data or examples that were used for the learning process known as model training. Consequently, the notion of transparency challenges this opacity by aiming to make the workings and rationale behind an AI model's decisions clear in a given situation [181]. While transparency might be seen as making the code available, its conceptualization is far more intricate. Opening the code might not illuminate the model's functioning due to its complexity, and even if it could, the number of people who would understand it remains limited. Thus, the concept of meaningful transparency [40] has emerged, emphasizing the provision of relevant and actionable information that is accessible to stakeholders according to their understanding and needs. Transparency extends beyond simply providing information about an AI system's outcomes; it also involves explaining the choices made throughout the stages of conceptualizing, designing, constructing, evaluating, and employing the AI system. For a user, transparency involves being notified when an AI system makes a significant decision or when they are interacting with one, obtaining useful information about how the AI makes decisions (relating to the concept of explainability that we discuss below), and accessing information about the data being utilized. More broadly, for stakeholders like policymakers, regulatory authorities, organizational leaders, or members of civil society, transparency means building and managing AI systems in a way that enables oversight. This includes raising awareness of the system's abilities and constraints and fostering open communication among all stakeholders, providing explanations for algorithmic decisions, justifying how the AI systems operate such as detailing the data used for training and how it was sampled or labelled, and ensuring traceability between these decisions and their outcomes [98]. Thus, beyond the connection to the explainability of AI outcomes, transparency is also crucial for facilitating the principle of human control over AI, albeit partially.

- **Explainability** refers to the ability to provide insights or reasons to make the operation of an AI system, especially its outcomes, clear or easy to comprehend to a target audience [13]. This notion is distinct from the interpretability of AI models, which denotes an inherent quality of an AI model that facilitates its comprehension [13], implying that rule-based models are far more interpretable than those derived from deep learning techniques. Explainability, instead, also covers the generation post-hoc information that helps elucidate an AI model, even those developed through deep learning, with a deliberate focus on the humans who need to grasp this information. Initial and significant efforts to enable explainability, such as LIME [250] and SHAP [190], are mainly targeted towards AI developers. They provide insights, such as identifying the most significant features influencing an AI decision, to assist developers assess the suitability of specific feature relationships for decision-making, whether for selecting an appropriate model or improving the best-performing models. Recently, the perspective has expanded to encompass human-centred explainable AI [91, 93], aiming to widen the range of AI explanation recipients and more thoroughly address their varied contexts, prior knowledge, and needs. Returning to the COMPAS scenario, adopting a human-centred perspective would prioritize the needs of the judge (the user) as well as the accused and their legal representative (those impacted by the AI's decision). This approach aims to develop explanations that enable them to comprehend the process behind the AI's conclusion, evaluate its suitability, and decide on further steps. These steps could vary from accepting the outcome and supplementing it with additional information to finding avenues to contest and rectify the AI-mediated decision. Therefore, explainability plays a vital role in fostering transparency and enabling human control of AI. Moreover, it is essential to evaluate the impact of AI's decisions on the ability to uphold human rights, assessing whether this is achieved equitably across diverse social groups. This ties into the principle of fairness, which we will explore in more detail below.
- **Fairness** arises as a countermeasure to algorithmic bias, a term denoting "a systematic deviation from equality that emerges in the outputs of an algorithm," [175] which could harm members of certain groups by, for instance, limiting their access to benefits or raising the likelihood of them facing penalties. ProPublica highlights the COMPAS system as an example of racial bias within judicial decisions, illustrating how false positive and negative rates vary considerably between races. This system tends to overestimate the recidivism risk for black defendants,

negatively impacting their bail prospects. Such disparities can be traced back to historical biases in law enforcement practices, like increased surveillance and incarceration rates within the U.S., from where the training data was collected. AI's bias is not confined to the justice system but spans across healthcare, education, employment, marketing, and more, exhibiting biases based on race, gender, and other factors. The principle of fairness, then, underscores the imperative to combat discrimination against individuals and groups, highlighting the elimination of unjust biases as essential to preventing social injustice and preserving the autonomy of individuals [167]. While the discussion on fairness in AI initially centred around unequal outcomes, it has expanded to urge designers and users to pay attention to biases and other factors that influence discrimination throughout the AI lifecycle [181]. Jobin et al. [157] emphasize the importance of not just preventing but also actively monitoring and mitigating biases and discrimination. Still, a significant focus is given to the critical role of data, which frequently contains ingrained societal biases from the past or lacks the inclusivity to reflect the diversity of those affected by AI systems. This situation has led to calls for the use of data that is both representative and high quality [101], considering aspects such as accuracy, consistency, and validity. Nonetheless, the issues of fairness extend well beyond the scope of data, encompassing the need to address biases, injustices, and discrimination throughout problem formulation, pre-processing, AI implementation, and evaluation of its outcomes [181]. Strategies to achieve this encompass involving a diverse array of stakeholders in problem formulation and design processes to ensure representation [98], integrating accessibility considerations and carrying out algorithmic audits, which relate to the principle of accountability. Overall, the goal is that people are treated fairly, devoid of discrimination based on race, gender, nationality, age, social class, political beliefs, religion, and disabilities, among others, with the broader objective of preventing any further disadvantage to marginalized groups and avoiding the continuation of historical biases. Additionally, there is an emphasis on the equitable distribution of the benefits and value generated by AI, ensuring that these advantages also reach those who are typically excluded or marginalized [101].

- **Privacy** is related to respecting the intimacy of people, their control over personal information, and their ability to make choices about their data and the decisions derived from it [277]. It encompasses aspects such obtaining consent to use personal data, allowing individuals to control how their data is used, providing the ability to restrict data usage, and granting rights to rectify and erase their data. Each of these factors plays a crucial role in reshaping how data is (and will be) gathered and subsequently utilized to train AI models. Moreover, they could impact the deployment of AI systems for activities that could infringe on individual privacy, like surveillance. This principle encompasses the development of measures to safeguard data against unauthorized access, which may involve techniques like data encryption and differential privacy [90]. Significant data breaches, such as the Cambridge Analytica case [150], and the re-identification of anonymized data in the Netflix contest for recommendations [215] and the US Census [286] are prominent examples of the vulnerabilities around data privacy nowadays. Unlike many other principles, a unique feature, in this case, is the existence of laws designed to safeguard data and privacy in most nations worldwide (in 137 countries out of 194 as of 2021 [227]). Among them, the European Union's General Data Protection Regulation (GDPR) [243] is playing a pivotal role in framing the debate on data regulation, the field of data science, and privacy expectations related to AI [101].

- **Accountability** underscores the importance of establishing mechanisms to ensure that responsibilities for AI systems' faulty outcomes and harmful impacts are appropriately assigned [101]. It comprises identifying both organizational and individual responsibilities in creating and implementing AI systems and it advocates for the establishment and adoption of legal frameworks to define and enforce these responsibilities. Special attention is given to scenarios where decisions are made without human intervention, alongside the significant impacts of AI on sustainability and society. Accountability also involves enabling the auditability of AI systems and having mechanisms to improve the systems after an audit. Moreover, it emphasizes putting strategies in place to assess and mitigate risks. It encompasses conducting impact assessments, verifying that the system functions correctly, giving enough information for third-party validation, establishing mechanisms to appeal AI's outcomes, and providing remedies for any harm AI may cause to the environment and individuals (e.g., compensations).
- **Safety and security** represent two critical dimensions of any information system, including those powered by AI. Safety ensures that the system operates according to its design without causing unintended harm, whereas security involves the system's ability to defend against unauthorized access or manipulation by external entities [101]. Safety requires the construction and verification of systems to ensure they operate as intended and prevent misuse, seeking to eliminate risks of harm. This encompasses the system's ability to function accurately, reliably and robustly, even under challenging circumstances. Crucially, it is essential to maintain safety's testing and monitoring practices

post-deployment, observing the AI systems' behavior as they learn from new data and adapt or, as they continue to work with the initial model parameters under different scenarios [181]. In turn, security is related to protecting the system and its components against adversarial threats. It seeks to ensure the system remains functional and accessible to legitimate users while protecting private information from unauthorized parties [181].

This categorization of principles, grounded in literature, covers a broad range of ethical concerns in AI. However, we acknowledge the dynamic and evolving nature of AI, which may give rise to new ethical considerations and challenges. Therefore, we view these principles not as final but as a starting point for ongoing dialogue, reflection, and adjustment. In the next section, we will present case studies to discuss the intertwined relationships among these principles and offer opportunities to reflect on the need of a sociotechnical perspective to address them more effectively.

1.7 Further readings

To learn more about applied ethics, we recommend Meynell and Paron's "Applied Ethics Primer" [202], Jackson et al.'s "Applied Ethics: An impartial introduction" [151], and Shafer-Landau's "Living with ethics" [270]. Ethical concerns in robotics are further explored in [23]. Hellstrom [138] and Galliot [109] address issues related to military robots, while Gunkel examines the possibility of assigning rights to robots [128]. Machine ethics is discussed in [114].

Other research addresses specific ethical issues surrounding algorithms and AI. We recommend classic books that discuss their impact on perpetuating gender, race, and social class inequalities, authored by Eubanks [97], O'Neil [228], Noble [218], and Benjamin [28]. Other researchers have examined the hidden and undervalued labor behind AI [124], as well as issues related to privacy [59].

Regarding AI principles, we believe that "Principled AI: A Map of Ethical and Rights-Based Approaches to Principles for AI" [101] from the Berkman Klein Center for Internet Society remains an excellent starting point for tracking efforts to define principles that can guide AI development. More specifically, key examples of soft and hard laws seeking to guide and regulate AI development include frameworks from the UN [142], UNESCO [295], OECD [65], the European Union [98], and an executive order from the White House [30]. A summary of attempts to regulate AI in Latin America, which includes the countries' declared AI principles, can be found in a technical policy report from Access Now [1].

Discussions about ethics washing as a way to avoid regulation are available in [304], and possible paths forward can be found in "From Ethics Washing to Ethics Bashing" by Bietti [31].

Chapter 2

A Sociotechnical Approach to Integrate Ethics into AI Projects

In this book, recognizing AI systems as sociotechnical systems will be the foundation for analyzing the ethical problems we encounter in AI and IT. In this chapter, we describe why we adopt a sociotechnical perspective to understand and study AI and provide examples of current perspectives that embrace it to integrate key ethical concerns into the development of AI projects.

2.1 What is sociotechnical AI?

This approach considers not only the technical aspects of developing this technology but also the intricate societal influences required to design, develop, implement, and use AI. In simple words, considering AI sociotechnical means acknowledging that AI does not operate in a vacuum but rather in complex contexts that demand attention to different aspects.

The notion of "sociotechnical systems" comes before its connection to AI. Researchers have characterized sociotechnical systems in different ways. Baxter and Sommerville [25], for example, talk about methods for designing sociotechnical systems as an approach that considers "human, social and organizational factors, as well as technical factors in the design of organizational systems. The outcome of applying these methods is a better understanding of how human, social and organizational factors affect the ways that work is done and technical systems are used" [25, p.4]. However, the basis for this definition can be traced back to the 1950s.

At the Tavistock Institute in London, the concept of "sociotechnical systems" arose as a response to several projects that were developing the British coal mining industry [292]. The post-war scenario meant that new technologies were being implemented, but also that interpersonal relations to manage the new labor dynamics needed to be addressed, including organizational arrangements to achieve comprehensive industrial development and productivity. In summary, the main proposal of this "new paradigm" is a shift in understanding and designing work organizations, emphasizing the transition from a purely technocratic approach to recognizing and attributing value to the importance of understanding the role of people in the institutional context. Hence, older paradigms about technological imperatives, where machines were at the centre, and humans were considered mere complements, were replaced with a notion that people and other societal dimensions are interconnected with technology.

Thus, sociotechnical research has been characterized as an endeavour based on bidirectional benefits that are derived from the intersection of social and technical elements [94]. Such intersection requires reciprocity between technologies and society, which shapes how both dimensions develop. In simple words, the success of a sociotechnical system can be evaluated based on the success of these interactions.

As the notion of "sociotechnical systems" evolved, so did the specificities allowing us to make deeper connections between societal and technical aspects of technologies, not just based at an organizational or management level. Pinch and Bijker [240], for example, follow the SCOT framework (Social Construction of Technology), which proposes that the development of technology is both a technical and social process, arguing that technological artifacts and systems are shaped by social, economic, and cultural factors, implying that technology does not evolve primarily through innovations of individual engineers or specific technological tools.

More specifically, Bijker [33] introduces new concepts that add depth to sociotechnical views. One example is the idea of interpretive flexibility. This concept suggests that a technological artifact can have different meanings and uses for different social groups. This means that the design, development, and use of technology are open to interpretation based on the users' needs, values, and social context. This flexibility allows for different technological solutions to the same problem. What Bijker is pointing out is that the meaning attributed to a technology or an artifact is not based on the technology itself. Furthermore, this distinction has an even deeper relation to what a sociotechnical system entails: "relevant social groups do not simply see different aspects of one artifact. The meaning given by a relevant social group actually constitutes the artifact" [33, p.77]. Hence, we cannot truly evolve and develop technological advancement without its sociotechnical understanding.

The concept of sociotechnical systems refers to systems that rely on a combination of technical infrastructure, human behavior, and social institutions to function effectively [76, 217]. More specifically, van de Poel [298, p.391] proposes that they comprise three fundamental components:

1. Technical artifacts, which are physical objects designed for specific technical functions, i.e., possessing a physical presence,
2. Human agents, individuals who carry out intentional actions and engage with technical artifacts, and
3. Institutions, which are social rules or norms that govern behavior, establishing expectations for moral conduct.

Over and above these three elements, AI systems have two additional unique components that influence their sociotechnical understanding. They are:

1. Artificial agents, and
2. Technical norms.

While human and institutional agents are typically understood in terms of intentionality, artificial agents and technical norms, says van de Poel, operate within a framework defined by causal or physical mechanisms. Artificial agents are distinct from conventional technical artifacts due to their capacity to exhibit or mimic human-like characteristics such as autonomy, interactivity, and adaptivity [103]. Unlike traditional technical artifacts, which primarily rely on physical structures, artificial agents have the ability to adapt to varying contexts and interact with other agents (artificial or not), constituting a more complex and dynamic behavior. But, as we know, although artificial agents can exhibit more dynamic characteristics than technical artifacts, their capabilities lack inherently human traits such as consciousness, free will, emotions, and moral autonomy.

Technical norms are dependent on the presence of artificial agents who do not follow traditional social rules, as these pertain to human interactions. Instead, technical norms serve as the AI equivalent of institutions, providing a set of rules and guidelines that govern the behavior of artificial agents. Technical norms are understood in causal terms, providing the operational boundaries within which artificial agents can function, ensuring consistency and establishing the constraints for artificial agents to operate, allowing them to function within AI systems without possessing human-like intentionality.

Hence, AI, understood as a sociotechnical system, means recognizing that there is a complex network involving interactions between technical elements, such as algorithms, artificial agents and technical norms, and social elements, such as human behavior and institutional aspects, including cultural norms and regulatory frameworks. A system's effectiveness and impact are shaped by these interactions, which operate in a dynamic feedback loop where technology influences, and is influenced by, human and societal practices, including ethical considerations and legal guidelines.

Adopting a sociotechnical perspective on AI highlights the need to understand AI systems in a broader context, considering not only their technical components but also their social consequences, which has a significant implication for developing AI ethics, i.e., designing AI systems that align with societal values, thus ensuring that technology serves the greater good. To achieve such a complex task, adopting a sociotechnical view of AI requires a collaborative approach involving ethicists, engineers, policymakers, and social scientists, among other professionals, to ensure that AI systems are designed and governed in ways that support ethical principles and societal values.

2.2 Key elements of sociotechnical approaches to AI

Now that you know what it means to talk about the sociotechnical perspective, we hope you can see why we believe this is fundamental to advancing AI ethics. The ways in which a sociotechnical standpoint can influence AI ethics are varied, and here, we will give you some examples highlighting key elements of sociotechnical approaches so that you can explore different ways in which this perspective can benefit your work on AI.

2.2.1 Value-based design of AI

Technical artifacts can embody values if they are designed with specific intentions that align with those values and if their use is conducive to achieving them. For instance, if a technical artifact is designed with safety features, and those features contribute to safety when used correctly, then the artifact embodies the value of safety [298]. The concept of embedding values in technical artifacts revolves around the idea that design choices reflect certain values that can also be understood as ethical choices. Additionally, these choices influence how the artifact functions in practice. The same happens to institutions that embody values through formal rules and practice standards, promoting specific outcomes and goals. However, as noted above, artificial agents do not follow this set of rules. Therefore, translating these values into their technical norms can allow us to guide AI's autonomous and adaptable behavior.

This translation of values into technical norms is not merely a supplementary feature but a fundamental aspect of ethical AI design, crucial for ensuring that AI technologies function in beneficial ways and are aligned with ethical principles. Incorporating ethical considerations into AI design, referred to as ethical design, mandates that AI systems be crafted not only with specific functional goals in mind but also with a keen awareness of their broader social impacts. For example, an AI tasked with content moderation on digital platforms must balance its operational efficacy with fairness and freedom of expression considerations. This balance can make us ensure that moderation does not unfairly target specific groups or stifle legitimate discourse, thus embodying the ethical values of fairness and respect for individual rights.

Overall, technical norms differ from social norms in that they are enforced through code and algorithms rather than social or moral expectations. Therefore, the call is to understand how human decisions about these technical features affect AI's sociotechnical context.

Furthermore, a key aspect of the sociotechnical perspective is recognizing that values can evolve as systems are used and adapted over time. It is not just the initial design that matters; the ongoing process of redesign and feedback is crucial to ensure that systems continue to embody the intended values as they evolve and adapt. Hence, continuous monitoring is required to maintain alignment with ethical principles and societal values as we learn from the interaction and sociotechnical symbiosis of AI and society.

Feedback mechanisms are vital for aligning AI systems with societal values. These mechanisms enable AI systems to adapt based on user and stakeholder input, reflecting an ongoing commitment to ethical congruence as societal values evolve. This dynamic adjustment process is critical as it allows AI systems to remain relevant and ethically aligned over time despite changes in social norms and values. The concept of the moral machine serves to illustrate this. For example, autonomous vehicles may be programmed to make decisions in morally charged situations, such as accident scenarios where harm cannot be completely avoided but can be minimized. The decision-making process in such cases is heavily laden with ethical considerations, balancing factors such as passengers' versus pedestrians' safety. Similarly, in healthcare, AI systems like robots in emergency rooms may need to prioritize patient care based on severity and urgency, embedding values of fairness and equity in medical treatment prioritization.

2.2.2 Adaptability: robust ethical governance

Chopra and Singh [56] argue for distinctions in moral decision-making within AI from a sociotechnical perspective. The authors criticize that traditional approaches to moral decision-making in AI often rely on decision-theoretic perspectives, meaning that they only focus on individual agents or on individual ethical problems. They claim this approach may be too narrow because moral decision-making is highly driven by context. Therefore, in contrast, they argue in favor of considering ethics from a broader sociotechnical perspective to achieve a more holistic analysis.

To develop their approach, they consider three elements of sociotechnical systems [56, pp.2–3] (1) stakeholders; (2) information, namely, stakeholder values, prescriptive norms, and outcomes, and (3) processes for governance, including for purposes of respecifying sociotechnical systems. Respecifying refers revising or redefining the parameters of a system, model, or project, and it is key because it involves interactive activities among stakeholders to ensure that the norms of a sociotechnical system align with ethical considerations over time.

For this, they specify three governance activities: design, enactment, and adaptation —ensuring that ethical governance of AI represents the dynamism provided by a sociotechnical lens. Design, for Chopra and Singh [56], involves creating a sociotechnical system that satisfies stakeholders' requirements, considering their values and engaging with them throughout the process. Enactment refers to the behavior within the sociotechnical systems, ensuring that they act as expected. Adaptation addresses the need for ongoing adjustments to the systems based on outcomes and changing requirements. This approach to governance promotes flexible and responsive systems that can evolve with the context.

Adaptability —a central theme of sociotechnical perspectives— reflects the need for systems to evolve in response to new information and changing norms. Researchers [56] highlight that norms may need to change to guide outcomes in a desirable direction. Additionally, computational mechanisms, such as algorithms, might need to adapt to meet new standards. This adaptability fosters innovation and allows for continuous improvement within the sociotechnical systems.

The implications of adopting a sociotechnical view of AI change the default of isolated or individual interventions and analysis of ethical problems and, instead, focus on a broader framework that is based on considering the context in which AI operates, focusing on governance and adaptability to ensure ethical outcomes. Hence, to develop ethical AI systems, we do not simply respond to individual decisions but rather explore how they interact with the different elements and stakeholders in the AI ecosystem to ensure more robust governance.

2.2.3 Interdependence: required multi- and interdisciplinary work

The interdependencies between technology and society are central to the discussion of AI as a sociotechnical system. Technological myths and narratives significantly impact the adoption and usage of AI technologies, influencing how developers, citizens, and policymakers perceive and engage with AI, as noted by Sartori and Theodorou [264]. The authors, emphasize the crucial importance of raising awareness among people, enabling them to critically adopt new technologies.

This consideration for awareness underscores another requirement: a multidisciplinary approach for incorporating perspectives from a wide range of stakeholders. For Sartori and Theodorou [264], adopting a sociotechnical view of AI systems requires us to "include all participants in the process of construction in a co-creation approach" (p.8). The narratives shaping the current realities and contexts in which AI is created reflect perceptions and beliefs influencing the future of AI. How these narratives can change across cultures is extremely relevant, as this can translate directly into the ethical principles we expect from AI and the decisions we make on how to develop it. Thus, the importance of interdisciplinary work in AI is highlighted by the need to understand and address the complex interdependencies between social and technical elements, which is supported by a sociotechnical view.

In practice, this can be translated into specific ways to integrate ethics by adopting sociotechnical views that have disciplinary contexts mediating the interdependence of technical and societal factors. For example, when integrating ethics into Computer Science education, Goetze [118] argues that interdisciplinary approaches are more effective than multidisciplinary ones because they closely connect ethical content with technical learning, allowing for a deeper understanding of ethical issues in computing. The author acknowledges that technical education is insufficient without incorporating ethical considerations. They suggest that students should be capable of analyzing and constructing ethical arguments as well as understanding the societal impacts of computing technologies. To achieve this, they need to be immersed in the context and ethical implications of their work, which can also benefit from a transdisciplinary approach. A deep integration of ethical and technical concepts can potentially lead to new paradigms in computing, with ethical success being as important as technical success. Another example is the analysis of Dolata et al. [83], where they argue for embedding algorithmic fairness in the sociotechnical view of information systems precisely to acknowledge and actively include the interdependencies of the social and the technical, because for them:

> "Without a coherent perspective that acknowledges the interdependencies between the social and the technical aspects of AI, organizations may be reluctant to effectively tackle this problem. If they treat algorithmic (un)fairness as a purely technical problem, they may assume that adding a social element will sufficiently solve unfairness." [83, p.765]

From a sociotechnical standpoint, the authors emphasize that fairness in AI is more than just reducing technical biases or achieving equitable outcomes. It involves considering how social structures, human interactions, and technical systems collectively influence algorithmic outcomes. A purely technical approach may overlook critical factors like cultural nuances, societal norms, and historical contexts, leading to unfair outcomes despite technical efforts to ensure fairness. Thus, a sociotechnical approach takes us away from traditional (technical) approaches, as Dolata et al. [83] call them, which are often focused on distributive justice, reducing fairness to a statistical problem. A sociotechnical approach considers other aspects of justice, such as interactional and procedural fairness, which are crucial in ensuring that AI systems are technically sound and socially just.

2.2.4 Situatedness: embodied context

Understanding the broader impact of AI on society is essential, and to achieve this, researchers have resorted to crucial concepts from feminist philosophers to emphasize the contextual requirements of a sociotechnical view. Early on in this discussion of integrating sociotechnical views in AI, Draude et al. [85] presented an approach of "situated algorithms". They argue that algorithms, often seen as mere technical entities, are increasingly being investigated as sociotechnical systems with implications for social inequalities and cultural hierarchies. A sociotechnical perspective, they stress, avoids a simplistic view that separates algorithms from their human context and considers how these systems interact with our social fabric. Thus it aligns with the idea of "algorithmic culture," where algorithms play a significant role in sorting and classifying, ultimately affecting our cultural landscape.

Hence, the authors use as a starting point Harding's criticism against the biases and power structures that historically shaped scientific knowledge. According to Harding [131], understanding knowledge production requires acknowledging whose perspectives are represented and who benefits from them. This approach challenges the idea of "value-neutral" objectivity, arguing that all knowledge production is influenced by broader sociopolitical contexts. In the AI context, the failure of facial recognition to accurately recognize faces of color, for example, reflects a deeper pattern of racial bias embedded in technological systems rooted in unequal representation in training data sets. Against this, Harding's concept of "strong objectivity", used by Draude et al. [85], proposes that to counteract power imbalances in knowledge production, research should start from the perspective of those most affected by inequalities. This stronger form of objectivity acknowledges its positionality and strives to interrogate existing power structures. Research endeavours, such as AI development that adopt strong objectivity begin by questioning their own role within systems of inequality and considering which groups might benefit or be disadvantaged. This approach aligns with the idea of situating algorithmic systems in their sociopolitical context, ensuring that design processes account for the embeddedness of algorithms in existing power hierarchies and cultural frameworks.

Overall, Draude et al. [85] argue that "to produce less biased and more accountable sociotechnical solutions, it is crucial to situate algorithmic systems and their design process, i.e. to understand and address their embeddedness in political, socio-cultural contexts and existing power structures." [85, p.335] Thus, for creating accountable and less biased algorithmic systems, they propose a systemic approach integrating the "4P" framework: People, Place, Power, and Participation. This framework addresses the broader impact of algorithmic systems by focusing on key questions about who is involved, who benefits, and who might be negatively affected.

2.3 Examples of sociotechnical approaches to AI

In this section, we want to highlight two approaches that have embraced the sociotechnical perspective and its key elements. These examples help us visualize how ethical concerns can be integrated into processes of AI development.

2.3.1 Data feminism

In connection to situatedness, D'Ignazio and Klein [81] proposed Data Feminism to emphasize lived experiences as essential sources to understand and shape data uses and limitations. This proposal invites us to understand data— a key input of machine learning— as necessarily reductive representations of some people's experiences, where lives are translated to specific numbers, words, or images, leaving many parts of their lived experiences unaccounted for. At the same time, data feminism invites us to note that others, usually more privileged, people play influential roles in counting, labeling, analyzing, using, and taking advantage of the value of the data. Both actors, those whose data is used and those who use the data, are parts of a context where multiple structures of oppression (i.e., by gender, race, class, ability, age, sexuality, geography and more) co-exist and compound. Thus, data feminism highlights the need to examine the context in which data was produced as well as the goals of its use from an intersectional point of view.

Intersectionality pays special attention to the compounded burdens of the experiences of people who live under multiple levels of oppression, such as women of color or migrants with disabilities from the Global South—which often go unnoticed by people who do not bear such burdens. Data feminism emphasizes the risks stemming from the significant gap between the lived experiences of those who make decisions about data-based systems—often individuals from homogeneous, privileged social groups (e.g., white, male, cisgender, nondisabled, educated in the Global North) and the experiences of those who are affected by these systems, who typically are more diverse populations. Acknowledging this gap is a necessary starting point to recognize the challenges in identifying how data-based systems might be causing harm or perpetuating biases.

Taking an intersectional point of view can also help us avoid misinterpretations of data as objective and neutral artifacts. Instead, it pushes us to understand data as a product of unequal social relations, where multiple actors and elements, such as people —including their goals and practices— available technologies and institutional norms can influence the characteristics, of the data. For example, in the COMPAS case, data feminism invites us to question why there is more data on black than white defendants and how that relates to historically broader surveillance and patrolling in neighbourhoods inhabited mostly by people of color. Thus, using data we recognize as heavily biased requires reflecting on its validity and limitations, questioning how well it represents the characteristics of people's experiences we want to measure, and deciding whether and how to use it as input for our AI projects.

The aim of highlighting people's experiences as data sources and decision-makers around data is to help us scrutinize how existing power imbalances influence the kinds of data-based technologies that are developed and deployed globally. Here, D'Ignazio and Klein propose examining whose goals are prioritized in data-based projects, who benefits from these projects, and whose lives are being "datified" and impacted by the outcomes. This includes considering the hidden and under-paid work of moderation and data labeling, as well as the environmental impact of natural resource extraction needed to sustain data and AI infrastructures. In this sense, data feminism seeks to question whether data-based systems contribute to keeping the status quo or reinforcing and even escalating current power imbalances.

Furthermore, it encourages us to use data and AI with a new goal: "to challenge power." This means using data to subvert current power asymmetries, seeking to respect the agency of vulnerable communities. A key idea in this respect is embracing pluralism for developing data-based systems. After acknowledging the reductionism of quantitative data, data feminism recommends coupling data-based systems with inclusive, participatory processes to inform their development and implementation with evidence-based insights from local, diverse perspectives. This pluralistic approach, connected with the interdependence dimension discussed above, must be complemented by mechanisms for knowledge transfer toward and from the affected communities and building the social infrastructure needed to sustain technical data-based interventions, thus ensuring adaptability in the long term. Other principles of data feminism include valuing emotion and embodiment over a neutral portrayal of data and visualizations, reevaluating binaries and hierarchies in available data, and making the labor behind data-based systems visible and appreciated.

2.3.2 Design justice

Costanza-Chock's Design Justice [64] uses the concept of situatedness to transform the design of technology-based solutions —a broader task than attending to specific problems around data or AI. The author critiques how the prevailing values, practices, narratives, sites and pedagogies in design reproduce systematic inequalities, maintaining current power distributions by, for example, tailoring technology affordances for more dominant, profitable social groups, and neglecting the needs of minorities, creating additional burdens on them. The work also scrutinizes how designers' practices and narratives hide and sometimes misappropriate contributions of users, erasing the collective, cumulative changes that lead to their breakthroughs. Thus, Design Justice helps us analyze the role of design in perpetuating social inequalities and calls on us to pursue a fairer distribution of design's benefits and burdens, aiming to achieve social justice.

Echoing themes from Data Feminism, Design Justice highlights the disconnect between those who design technologies and the communities they aim to serve. Drawing on disability activism and its mantra, "nothing about us without us," the author makes a strong case in favor of the involvement of community members who are directly impacted by technology in the design process. Importantly, this involvement must be substantive and continuous throughout the design lifecycle from its very beginning. Unlike other participatory design approaches, Design Justice advocates for the active participation of community members in defining the scope of projects and framing the problems, ensuring that their lived experiences significantly influence the definition of the issues to be addressed by the design process. This means that participation extends beyond simply brainstorming ideas and assisting the design team with prototype testing. It involves having a say in determining which problems need to be addressed and how they should be tackled. This approach is advocated for reasons of justice, but it also has practical benefits: the unique insights, lived experiences, and tacit knowledge of community members can lead to innovative ideas and perspectives that might otherwise remain undiscovered by those outside the community. Involving community members in framing problems can also shift us away from technocentric solutionism when addressing complex and nuanced societal issues.

Design Justice also advocates for diverse design teams that include community members but also work towards transforming traditionally extractive design methods into mechanisms that allow communities to receive credit, visibility, profit, and ownership of the designed artifacts as retribution for their contributions to the design process. This transformation not only promotes justice as an integrated design value but can also help ensure the long-term adaptability of the designed technology. Furthermore, Design Justice emphasizes the importance of integrating intersectionality into the artifacts used to design and assess technologies. Constanza-Chock argues that we "need to develop intersectional user stories, testing approaches, training data, benchmarks, standards, validation processes, and impact assessments, among many other tools." [64].

Thus, Data Feminism and Design Justice offer a set of insights and practical suggestions that operationalize some aspects of the sociotechnical perspective to analyze and build AI projects, integrating contextualised ethical concerns into the practices of data work and design. Before moving onto ways of tackling ethical concerns in specific stages of an AI project, the next chapter will address specific ethical issues that have remained somewhat hidden from the mainstream debates in AI Ethics and need more attention as we further develop this field.

2.4 Further readings

To gain further insight into sociotechnical systems, we invite you to read the seminal paper on the topic [95], Trist's "The evolution of sociotechnical systems" [292], and Baxter and Sommerville's framework for designing and engineering sociotechnical systems [25]. For a deeper understanding of AI from a sociotechnical perspective, we suggest van de Poel's work [298] and Draude's research on situated algorithms [85] which are highly valuable. Wiggins and Jones [313] provide a sociotechnical historical perspective on the development of data and data-based decision-making. In general, sociotechnical systems are thoroughly examined within the field of Science, Technology, and Society. Therefore, we encourage readers to explore such literature for more detailed insights into sociotechnical approaches and critiques. For instance, Bijker's book [33] presents a series of case studies to explore and theorize key concepts within the sociotechnical perspective.

We strongly recommend D'Ignazio and Klein's Data Feminism [81] and Costanza-Chock's Design Justice [64] to find frameworks that embody sociotechnical elements to re-imagine ways to build AI to fight social inequalities. Additionally, Klein and D'Ignazio have written an article specifically addressing data feminism in the context of AI [171]. Beyond algorithms and AI, Criado-Perez [68] presents a range of compelling examples illustrating how data biases have historically led to technology developments that favor men over women. This highlights the broader need to create technology that intentionally addresses and reduces inequalities rather than inadvertently (or by omission) perpetuating them. These works draw on feminist perspectives in science and technology. In this area, we underscore the contributions of Harding's research [132, 133].

Chapter 3

Beyond the Mainstream: Sustainability and the Replicability Crisis

In the previous chapters, we introduced and exemplified core principles in AI ethics and a sociotechnical approach to complement them and tackle different ethical concerns arising from the design, use, and implementation of AI systems. These topics can be categorized under "the mainstream" discussions in the field of AI Ethics. In this chapter, we want to examine two issues that have been out of the mainstream and have only recently received attention: sustainability and the replicability crisis. We strongly believe that these topics require further attention as we continue to develop the field of AI ethics.

3.1 Sustainability

Van Wynsberghe [299] can be recognized as one of the first researchers to discuss the conceptualization of sustainability in AI. For her, this discussion constitutes the third wave in AI ethics. She characterizes the first wave as one focused on the risks of superintelligence and robot uprisings. The second wave is mainly what we have discussed so far in this book, addressing ethical concerns about machine learning and linking them to practice, with problems such as biases, discrimination, explainability, and privacy, among others. But the third wave, instead, "confronts the environmental disaster of our time head-on and actively seeks to engage academics, policymakers, AI developers and the general public with the environmental impact of AI" [299, p.213]—thus placing sustainability as a core element.

To develop her conceptualization of sustainability, Van Wynsberghe makes a crucial distinction between AI for sustainability, i.e. towards sustainable development goals, and the sustainability of developing and using AI systems. The first sustainability branch is probably the most well-known, as it seeks to apply AI to achieve sustainable practices and outcomes. For example, machine learning can be used to optimize processes for creating clean energy or drinking water. The second one, and perhaps the one that creates more ethical concern, has to do with fostering change in the entirety of the AI developing cycle from idea generation to governance. Accordingly, it is a notion of sustainability that aims "towards greater ecological integrity and social justice." [299, p.213]

Based on this foundational distinction, van Wynsberghe introduces Sustainable AI through the assessment of the whole sociotechnical AI system. This involves not only the application of AI technologies but also the critical need to address the broader sociotechnical context within which these technologies operate. The proposed definition points towards a movement aimed at redefining how AI technologies are developed and used, ensuring they are compatible with sustaining environmental resources and societal values across current and future generations.

Sustainable AI, from a sociotechnical perspective, integrates ethical considerations and sustainability across the entire ecosystem that supports AI development. Van Wynsberghe emphasizes the substantial environmental costs associated with AI, particularly the energy-intensive nature of training deep learning models. For example, she references a study by Strubell et al. [283] where they showed that training one NLP model can lead to an estimated carbon footprint of 626,155 CO2e (lbs) in comparison with common consumption rates for other technologies like car use in 1 lifetime of 126,000 CO2e (lbs). She points out that the carbon footprint of these activities are significant. The emissions from training a single AI model is comparable to the emissions from five cars over their entire lifetimes. This stark comparison highlights the urgent need for the AI research community to consider the ecological impacts of their work, not just the technological advancements and the ethical implications in terms of discrimination, privacy, or other moral demands.

Moreover, she calls for increased accountability within the AI industry concerning its environmental impact. Van Wynsberghe's call to action includes directing funding towards more sustainable AI methodologies, thereby incentivizing research and development that prioritize reduced energy consumption and lower carbon emissions. This redirection of resources is posited as essential for fostering innovations that align with the goals of reducing global carbon footprints and achieving sustainable development. However, achieving this is a challenging task, and further efforts need to be made to integrate these ethical imperatives into practical decision-making, balancing innovation and technological development with sustainability standards.

Another researcher who has investigated these concerns is Crawford [67] in her book Atlas of AI. She stresses the importance of key concepts that can alter or manipulate the public's perception of AI: "Advanced computation is rarely considered in terms of carbon footprints, fossil fuels, and pollution; metaphors like "the cloud" imply something floating and delicate within a natural, green industry" [67, p.34]. Furthermore, in the chapter "Earth", Crawford argues that the production and deployment of AI are intrinsically linked to significant ecological degradation and resource depletion. She meticulously details the extraction of minerals necessary for electronic components used to create the hardware necessary to train AI models. This extractive requirement not only causes environmental damage, such as deforestation and water pollution, but also raises profound ethical concerns about the sustainability of AI technologies —in line with the concerns raised by van Wynsberghe.

Crawford points out that the fact that a vast majority of the resources needed to power AI systems come from non-renewable energy sources directly contradicts the technology's perceived efficiency and beneficence, laying bare the environmental costs obscured by the industry's rapid growth. Her analysis also extends to the resultant electronic waste from obsolete hardware, which accumulates toxins in landfills primarily located in less economically developed countries. Hence, the socio-economic and environmental injustices perpetuated by the AI industry also come from unsustainable practices. Crawford argues that the burdens of AI development—ranging from labor exploitation in mineral extraction to the indiscriminate disposal of e-waste—are disproportionately affecting the Global South. This geographic disparity in how the environmental costs of AI are distributed raises significant ethical issues that are often missed by the "mainstream" discussion of the AI ethics second wave.

3.2 Replicability crisis in AI

At the end of 2023, Phillip Ball [327] commented in Nature that the naive application of AI is potentially contributing to a reproducibility crisis across various disciplines. In his article, he highlights the work of various researchers who have shown how AI systems have failed. Ball claims this is due, at least in part, to the improper use and widespread misunderstanding of AI and machine learning tools, especially in their training and testing phases. This discussion highlights two crucial aspects: the innovative use of AI technologies in new contexts and the potential pitfalls of deploying AI without rigorous validation.

An example of these pitfalls is the case of a machine learning system used to analyze X-ray images to detect COVID-19 during the pandemic. As the pandemic advanced, testing kits were scarce; therefore, researchers in India proposed that they could train a machine learning model to learn from chest X-ray scans to discern differences between infected and non-infected patients [168]. The paper was cited several times; however, almost a year after its publication, Dhar and Shamir [78] trained a similar algorithm using a fraction of the same images with blank backgrounds only, that is, not showing any body parts. Surprisingly, the re-trained algorithm was also capable of "detecting" COVID-19 cases.

The problem seemed to be that there were consistent differences in the backgrounds of the medical images in the dataset. An AI system could pick up on those artifacts to succeed in the diagnostic task without learning any clinically relevant features —making it medically useless. Their investigation revealed that the AI could identify COVID-19 from parts of the X-ray images that contained no diagnostic information, merely background noise. This discovery points to a fundamental issue in machine learning models: the risk of generating misleading results due to unrecognized biases or anomalies in the training data.

Another issue pointed out in Ball's analysis is the case of data leakage, which occurs when information extraneous is erroneously incorporated during a model training process. This phenomenon can occur when data intended solely for the test set, which serves to evaluate the model, is inadvertently included in the training dataset. Data leakage can also happen when preprocessing steps, such row duplication for oversampling, are applied to the entire dataset instead of being confined to the training subset. Consequently, data leakage results in an overestimation of the model's performance, as the model seemingly excels on data that it should not have been exposed to during its training phase.

Kapoor and Narayanan [164] highlight significant issues related to data leakage in machine learning research across 17 disciplines. In their study, they reveal that data leakage leads to overoptimistic assessments of model performance and irreproducible results, contributing to the reproducibility crisis. The authors provide a detailed taxonomy of data leakage, distinguishing eight specific types, ranging from simple mistakes in data handling to complex issues that require further research to fully understand and mitigate. To combat these issues, Kapoor and Narayanan introduce model information sheets designed to help researchers systematically document and verify that their models are free from data leakage. Thus, the authors argue, validating AI models against robust standards can contribute positively to society, but the uncritical acceptance of AI-generated outcomes without thorough scrutiny, particularly in areas of high social impact, is worrisome.

Complementing previous criticisms, Bausell [24] articulates that the reproducibility crisis in scientific research should not be perceived merely as isolated incidents but as a manifestation of systemic failures. He argues that these failures are rooted in flawed research methodologies and a prevailing culture that disproportionately values sensational results over meticulous and rigorous investigation. This aspect is particularly pertinent in the context of AI, where the allure and novelty of employing machine learning methods often overshadow stringent methodological rigour.

Furthermore, Bausell discusses how publication bias—where journals exhibit a preference for publishing positive results over negative or null results—contributes to a distorted scientific record. This bias is increasingly evident in research related to or utilizing AI, perpetuating the cycle of hype surrounding emerging technologies. Similarly, Bausell highlights questionable research practices, such as selective reporting and optional stopping, which he recognizes are not mere lapses in judgment but are frequently the product of institutional pressures and incentives. These practices are deeply entrenched in some research environments, leading to their passive acceptance or oversight. This scenario fosters a lack of rigour in AI research and its integration across various scientific domains, thereby exacerbating the existing challenges in scientific reproducibility. Bausell's critique underscores the need for a cultural and methodological shift within the scientific community to address these systemic issues effectively.

Assessing the replicability crisis in science, more generally, also has a philosophical dimension that relates to the ethical standards we need to enforce to avoid the mentioned crisis. Romero [255] analyzes this issue from a Philosophy of Science perspective. He discusses three critical solutions to the replicability crisis, classifying them into statistical reforms, methodological reforms, and social reforms. Each category addresses different aspects of the systemic flaws that contribute to the crisis, emphasizing the integration of research ethics and social epistemology views to tackle the replicability crisis.

1. **Statistical Reforms:** These reforms call for a fundamental shift in how data is analyzed within scientific research. Proposals include adopting Bayesian statistics, which demand transparent assumptions and offer direct inference of null hypotheses, thereby aiding in addressing replication failures. Another significant proposal is to lower the p-value threshold from 0.05 to 0.005, aiming to reduce false positives and enhance the statistical rigour of published studies. Additionally, emphasizing effect size and confidence intervals over mere statistical significance could shift focus towards the practical implications of research findings, enhancing their applicability and reliability.
2. **Methodological Reforms:** These reforms target the procedures of conducting and reporting research. Preregistration of studies is promoted to curb selective reporting and p-hacking by committing researchers to their initial plans before data collection begins. The adoption of open

science practices, including the sharing of data and materials, aims to increase transparency and facilitate the verification and replication of scientific work. Registered reports, a novel publication format, involve peer review prior to results being known; this approach reduces publication bias as acceptance is based on the research question and methodological rigour, not the novelty or significance of results.

3. **Social Reforms:** Addressing the cultural and institutional factors that discourage replication, these reforms suggest overhauling the academic reward system to recognize and incentivize replication efforts. Allocating specific funding for replication studies and adjusting academic tenure and promotion criteria to value replication and methodological rigour over the publication of novel findings are strategies intended to realign scientific incentives with practices that foster robust and reliable research outcomes.

In a more optimistic fashion, remaining in the same line of argument, Munafò et al. [212] argue that the reproducibility debate should be seen as an opportunity rather than a crisis because it highlights areas within the scientific research framework that can be improved for future robustness and integrity. This perspective stems from the potential to implement systemic changes that enhance the quality and credibility of research outcomes. This forward-thinking and proactive approach sheds light on constructive development rather than dwelling on the shortcomings of current practices.

Moreover, in a report by the OCDE [223], there is a section dedicated to the improvement of reproducibility in AI research, aiming to increase trust and productivity. There, Gundersen points out that almost 70% of AI research may not be reproducible, emphasizing that sources of irreproducibility include the design of studies, choices of machine learning algorithms, data handling processes, and the evaluation and reporting of research findings. These factors can lead to significant disparities in the replication across different environments or when different methodologies are applied, for which Gundersen proposes a series of recommendations to enhance the reproducibility of AI research. The author advises that research institutions enforce best AI research practices, including thorough training and quality assurance processes. Publishers are encouraged to standardize their review processes and to mandate the publication of code and data alongside research findings to facilitate verification. Funding agencies are recommended to prioritize transparency and open research by mandating that the research they fund be published in open-access formats and that all research outputs, including code and data, be freely shared.

As illuminated by several scholars, the discourse surrounding the reproducibility crisis in AI research not only exposes inherent challenges but also stresses an urgent need for ethical considerations. The naïve application of AI in fields such as medical diagnostics without rigorous validation leads to misleading results. This case underscores the critical ethical implications of depending on AI systems that may base diagnostic outcomes on irrelevant data features, thereby potentially endangering lives due to false medical assessments. Such incidents highlight a profound responsibility to ensure that AI systems are not only technically proficient but also ethically developed to prevent harm and promote trust.

In addressing these issues, the broader scientific community, including entities like the UK Reproducibility Network and OECD [223], advocate for multidimensional reforms spanning statistical, methodological, and social domains. These reforms aim to recalibrate the scientific paradigm by prioritizing ethical standards and robust research practices over sensationalism and novelty. The emphasis on ethical AI deployment, rigorous validation processes, and the transparency of research practices not only enhances the reproducibility and reliability of scientific outputs but also ensures that AI advancements contribute positively to societal welfare. This holistic approach to tackling the reproducibility crisis in AI research underscores the indispensable role of ethics in guiding technological progress, ensuring that AI serves as a benevolent tool of scientific inquiry.

3.3 Further readings

A pivotal book on the environmental impact of AI is Crawford's The Atlas of AI [67]. For insight into sustainable AI, see van Wynsberghe's work [299] as well as [211] for further connections between sustainability and environmental ethics, where Moyano-Fernández and Rueda discuss both AI for sustainability and the sustainability of AI, considering both operational and embodied carbon footprints.

Other recommended readings include Bossert and Hagendorff's work [38], which critiques human-centric approaches to sustainability, emphasizes future human generations, and advocates for considering animal welfare. Additionally, Erik Gundersen's work explores the fundamental principles of reproducibility, clarifies the differences between replicability and reproducibility, and discusses achieving a balance between transparency and scientific rigour [126, 127].

II

FAIRNESS, ACCOUNTABILITY, AND TRANSPARENCY

Chapter 4

Bias in AI

4.1 Introduction

The rise of AI and its myriad applications in everyday life have spurred efforts to better understand the potential risks associated with its use [273]. Its influence on our daily lives is set to increase significantly. AI assists us daily in making decisions [63], from choosing a movie and translating a text to hiring staff for a company. Its use has clear benefits, primarily triggered by the capability of inductive learning to analyze large volumes of data and extract subtle statistical patterns that would be impossible to detect with human effort alone. AI saves us time by efficiently processing large data volumes and identifying relevant information that benefits us.

The numerous advantages of AI heavily depend on the machine learning algorithms it relies on. Machine learning is a subfield of AI focused on developing inductive learning algorithms. These algorithms use data to identify patterns, which are encoded by the algorithms into models. For example, classifiers generate models to infer a target categorical variable, known as a target variable, for different types of objects. For instance, a system that recognizes digits (Optical Character Recognition) relies on classifying images of these digits from which it learns data regularities. These regularities correlate with the target variable, and the classifier can generalize to new instances from these associations. It is said that the model learns because it develops the capacity to infer the class of objects that were not used during the classifier's training.

As machine learning's basis lies in learning associations between descriptors of the objects it works with, it is pertinent to ask what these associations encode [57]. The risk is that these associations could solidify biases that affect individuals, groups, or subgroups within society, and therefore, a decision based on one of these systems could eventually be unfair [63].

Although there are various definitions of fairness in AI, and many stem from a robust tradition in other areas such as philosophy and social sciences, the basic essence of all these definitions is the absence of any prejudice or unjustified favoritism toward an individual or group based on their inherent characteristics [107]. Therefore, an unfair machine learning model can be the result of relying on biased regularities toward specific groups or individuals [147]. We will illustrate this situation with two examples. First, with an AI system designed for the US judicial system, which we have already introduced in Chapter 1 followed by a beauty contest where AI was an official judge.

4.1.1 Revisiting the COMPAS software

The Correctional Offender Management Profiling for Alternative Sanctions (COMPAS) [219] is a case management software developed by Northpointe, Inc. (now Equivant). Courts in New York, Wisconsin, California, and Florida used it. It was designed to assess the likelihood of an individual re-offending. COMPAS assigns risk scores based on human behavior descriptors.

Its development was motivated by a desire to minimize cognitive biases and prejudiced influences in judicial decision-making, one of the most noted being the "hungry judge effect." This effect highlights how judges tend to be more lenient after lunch and stricter before it. A related finding showed that more complex cases are scheduled for the morning, while simpler cases are handled in the afternoon due to their expected duration.

In July 2016, the Wisconsin Supreme Court ruled that COMPAS risk scores could be considered during sentencing. However, criticisms quickly emerged. A primary concern was that, as proprietary software, COMPAS is algorithmically opaque and cannot be publicly audited or examined. This information asymmetry leaves the analyzed individuals at a disadvantage, as they lack access to the underlying data supporting their sentences, complicating their ability to appeal and their right to an explanation (something now explicit in the EU's AI Act).[1] The foundation of such sentences could be addressed using a "white box" model —an open system that can be inspected. However, both opaque and open systems based on machine learning face challenges since the algorithms are data-dependent.

[1]Although explanations are not a guarantee for people to understand the reasons or logic behind the algorithm's decision-making, the debate for more interpretability and explainability is closely linked to the idea that users have a right to be informed about the due process of their information or at least offered an understandable justification for the outcome affecting them.

ProPublica's investigation into COMPAS [9] found biases against Black defendants. The study revealed that people of color were nearly twice as likely as white people to be labelled high-risk by COMPAS, yet many of these supposedly high-risk individuals did not commit violent crimes again. Conversely, white defendants were almost twice as likely as Black individuals to be labelled low-risk, yet a significant number of these cases did reoffend. The predicted trends, identified by the algorithm from historical data, can be associated with well-documented patterns of discrimination against Black people in surveillance, detention, and imprisonment in the USA, where the data was collected.

4.1.2 A beauty contest

Another example of biases in AI was found in Beauty.AI [26], which introduced the First International Beauty Contest judged by AI. Participants downloaded an app to submit selfies, which were evaluated by algorithms to determine the winners, who would be promoted in various global news outlets.

After announcing the winners in 2016, controversy erupted. The results highlighted a significant issue: in addition to using factors like facial proportions, symmetry, and wrinkles to judge attractiveness, it appeared that the AI disfavored individuals with dark skin. Among 6,000 participants from over 100 countries, including India and Africa, the top 40 spots were predominantly occupied by Caucasians, with only one top-40 contestant having dark skin.

While the evidence from the beauty contest might be considered anecdotal, the controversy underscored the critical impact of biases in algorithmic outputs. Depending on the context in which they are used, the consequences of these biases can be devastating [71].

4.1.3 The need to address biases in AI

Algorithmic unfairness can arise from multiple sources, notably biases stemming from data and those originating from algorithms themselves [284]. Despite AI benefits, implementing these systems entails significant responsibility. This responsibility involves addressing the impacts of algorithmic bias and properly addressing the harmful effects it may produce in society. The impact of algorithmic bias can be mitigated by early identification of its sources, thereby allowing for the reduction of its effects [285].

4.2 Biases in AI

Defining bias is not an easy task. In statistical contexts, bias could merely mean an inaccurate behavior of an AI system. In legal contexts, it focuses on aspects of disparate impact. In cognitive and social contexts, human decision-making is a more relevant influence. Here, we do not specify much about the contexts of bias interpretation but rather consider contextual influence concerning some established biased definitions. We recognize at least three properties to characterize bias in AI following the work of Zhai and Krajcik [321].

First, bias implies an error, in other words, a deviation between observations and ground truth. Biases also need a systematic component; they are systematic errors and not just random instances. Moreover, when we refer to bias, we also allude to an unjustified or irrelevant tendency in favor or against some ideas or entities over others.

Now, agreeing that bias is a systematic error with an underlying tendency of prejudice, its origin can be diverse, and we recognize at least three contexts of origin [267, 11]: societal, technical, and cognitive.

4.2.1 Societal, technical, and cognitive biases

Societal biases in AI stem from the social, cultural, and institutional contexts in which data is generated and used. These biases reflect and perpetuate existing prejudices and inequalities present within society.

AI systems often learn from large datasets that are generated by humans. If these datasets contain societal biases—such as racial, gender, or economic disparities—the AI system is likely to capture and replicate these biases. These biases embedded in data reflect historical and cultural prejudices, leading to discriminatory outcomes when AI systems are deployed. For example, an AI system used in hiring might favor certain demographic groups over others if trained on data that reflects biased hiring practices. Therefore, we refer to societal biases as systematic historical, institutional, and social influences, which reflect existing discriminatory behaviors, assumptions, and structural inequalities.

Technical biases arise from the specific methods, processes, and tools used to develop and deploy AI systems. These biases are linked to algorithmic constraints and developers' technical decisions, such as data selection and model design.

When data used to train AI models does not represent the entire population, biases can occur. For example, using data from a single, homogeneous source may introduce biases affecting AI performance across diverse scenarios. The technical design of an AI system, including the selection of features, model architecture, and optimization criteria, can introduce biases if not carefully considered. Technical biases can also result from errors or limitations in the algorithms, such as overfitting certain patterns in the training data or failing to generalize to new contexts. Hence, by technical biases, we mean errors in representation or systematic statistical errors that involve a certain level of partiality or discriminatory consequences.

Finally, cognitive biases refer to human errors in judgement or reasoning that can affect the functioning and deployment of AI systems. These biases occur when AI systems reflect the cognitive limitations and heuristics of their developers and users, when they mimic human decision-making patterns, or when humans make biased methodological decisions that contribute to the existence of technical biases, such as selection bias (see section 4.2.2).

AI systems often learn from interactions with humans, which can introduce cognitive biases if the humans involved in developing or using the system hold certain erroneous beliefs or make flawed judgements. Training data that includes human decision-making patterns can embed cognitive biases in AI systems. If humans consistently make biased decisions in certain contexts, AI trained on this data may replicate those biases. For example, AI can learn to "objectify" women's bodies if people who labelled images of bodies of women considered them more sexually suggestive than images of men's bodies [289].

Another example can be found in well-known medical data biases stemming from professional biases in certain medical specialities. Professional biases in AI systems can emerge when the systems are trained on data that reflects the cognitive biases of the professionals involved in their development or use. These biases can be inadvertently embedded in AI models when human decision-making patterns are incorporated into the training data. For instance, if healthcare professionals have a consistent bias in diagnosing certain conditions more frequently in specific demographic groups, an AI system trained on such data may replicate and even reinforce these biases. For example, racial biases relate to medical practitioners overlooking diagnostic features, causing certain non-white groups to be underdiagnosed or misdiagnosed compared to white patients [18], as well as having different levels of access to healthcare [14]. This may be attributed to biases that lead healthcare providers to downplay or ignore the symptoms reported by these groups as well as historical bias against women's pain [262]. Women, especially in gynaecological contexts, often report that their pain is dismissed or downplayed by healthcare providers. Conditions like endometriosis, polycystic ovary syndrome, or chronic pelvic pain are frequently underdiagnosed or misdiagnosed because women's pain complaints are not taken as seriously as they should be [268].

Thus, AI systems can reflect cognitive biases, especially if developers and people who generate the data unknowingly encode these biases into the systems. Therefore, with cognitive biases, we broadly refer to systematic human errors related to implicit biases and heuristics and the consequent influence of human decisions on AI's developing processes.

With this distinction made, we can now more closely examine some of the most common biases in machine learning. They include those that arise from the data collection (**data bias**) and those that are introduced and exacerbated during the training phase (**algorithmic bias**) [300]. Figure 4.1 shows a simplified machine learning process and its key inputs (data) and outputs (models). We use this Figure to illustrate where biases emerge in machine learning.

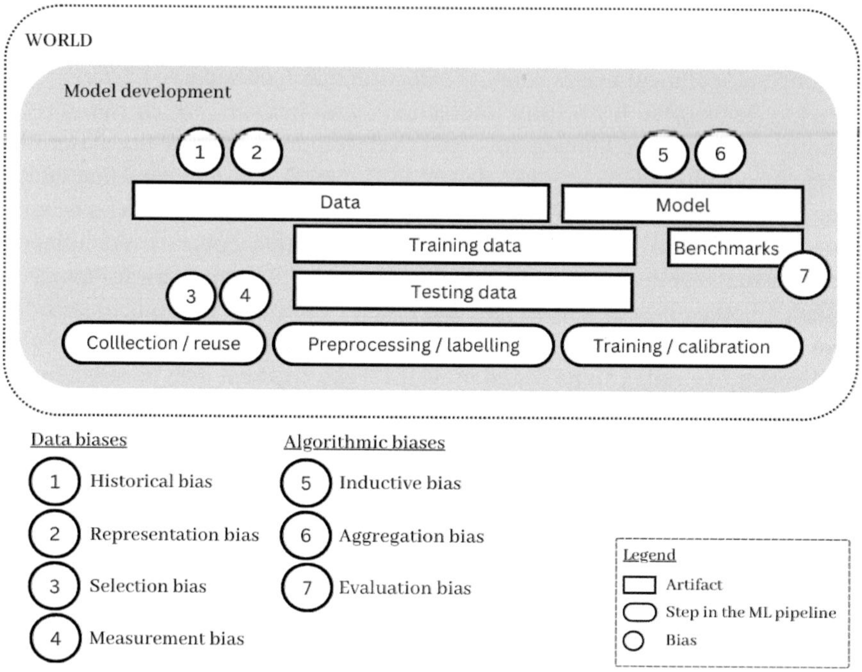

Figure 4.1: Examples of biases that may affect a machine learning model.

4.2.2 Data biases

Before exploring the specific types of data biases, it is important to recognize that biases in data can originate from any of the three primary sources: societal, technical, and cognitive. These biases can significantly influence the outcomes of machine learning models, leading to skewed or unfair results [22]. Data collection, which precedes exploratory data analysis and model training, is a critical phase where such biases can be introduced. During this phase, a target population is identified, and typically, a sample is drawn from this population due to the impracticality and costs associated with collecting census-level data.

There are various types of data biases, as shown in Figure 4.1. We will discuss them to understand their nature.

Historical bias

Historical bias in data reflects entrenched patterns of structural inequalities and discrimination that have developed over time [52, 195]. This bias can manifest in various forms, such as increased surveillance in predominantly Black neighbourhoods in the USA, beauty standards favoring Caucasian characteristics over those of other races, or the persistent gender wage gap in the labor market.

Historical bias arises from the ingrained patterns of inequality, discrimination, or stereotypes. These have been held over time within a society and find their way into AI systems, mainly through training data. This bias is embedded into the data collected from historical contexts, reflecting the social norms, practices, and power dynamics that existed when the data was gathered. Unlike other forms of bias that may result from flawed experimental design or sampling errors, historical bias occurs even when the data collection process is methodologically sound [300]. Hence, the key issue with historical bias is that it mirrors the world as it was rather than "how it should be". This can perpetuate past injustices and inequalities when this biased data is used to train machine learning models that are used to make or inform decisions about the future.

The data collection process is a critical stage where such biases can be introduced. For instance, in NLP systems, word representations are constructed through pre-training on large volumes of text, creating word embeddings. These texts are often sourced from publicly available repositories like Wikipedia, Google News, and Book Corpus. Since these sources mirror the societal context at the time of their creation, they inevitably reproduce the stereotypes and biases prevalent in that era.

Research has demonstrated that gender stereotypes, for example, can lead to problematic associations such as "man is to doctor as woman is to nurse" [35]. While this might appear anecdotal, the implications are far-reaching, as systems based on these biased representations can cause significant harm. As highlighted in Noble's work [218], search engines have been shown to perpetuate stereotypical representations of Black individuals. This occurs because text representation in NLP systems fundamentally relies on the spatial relationships between word embeddings in a latent space. Word analogies, therefore, decode relationships that the machine has learned from biased data and use these relationships to generate results.

Today, NLP-based applications are widely integrated into various systems that people interact with daily, such as chatbots, automatic translation systems, speech recognition, and web search engines. Consequently, these biased representations are already influencing how users receive and interpret results from numerous AI-mediated services, underscoring the critical need to address historical biases in data.

Representation bias

Representation bias occurs when the dataset used to train a machine learning model fails to accurately capture the full diversity and variability of the target population, leading to poor generalization for certain subsets of the population.

This bias can emerge in several ways and can result in skewed or inequitable outcomes, where the model performs well for over-represented groups. However, it under-performs or makes erroneous predictions for underrepresented groups.

As highlighted by Suresh and Guttag [284], representation bias can arise during the definition of the target population if it does not adequately reflect the intended use population. For instance, data representative of Brazil may not be generalized to the population of Japan, or data from Hanover, Germany, 30 years ago, may not accurately represent the current population. Similarly, if the target population contains certain naturally underrepresented groups, such as pregnant individuals within a medical dataset of adults, the model may be less robust for these minority groups due to the limited data available about them.

Moreover, representation bias can occur during the sampling process if the method used is limited or uneven, resulting in a development sample that represents a skewed subset of the target population [284]. For example, in modelling an infectious disease, if medical data is only available for individuals who were deemed serious enough for further screening, the resulting sample would be biased towards more severe cases, leading to a model that may not perform well across the entire population.

In some cases, the data may accurately represent the target population yet still exhibit representation bias. This happens because if a descriptive variable of groups is relevant to the task, machine learning algorithms require a balanced sample regarding this descriptor. It is then recognized that class imbalance is problematic in training classifiers, as an unbalanced sample based on the target variable increases the risk of over-fitting to the majority classes.

Furthermore, representation bias may arise due to the effects of transfer learning. If a model trained on a target population is then applied to a different population, it will replicate the stereotypes of the target population onto the new one. For instance, in the problem of bot detection, widely used systems on Twitter (now X) like Botometer [75], which were trained on data from an Anglo-Saxon population, yield inconsistent results when applied to a Hispanic population [197]. They tend to over-represent the actual volume of likely bots in Spanish-speaking countries. This discrepancy occurs because transferring a model from one linguistic and cultural context to another neglects cultural differences, local idiomatic uses, and specific behavioral patterns of the new context that were not observed in the target population. In this case, representation bias is caused by the under-representation of the new context's unique cultural and linguistic differences.

Hence, representation bias arises from methodological deficiencies during the sample collection process [271]. This type of bias is associated with an unfair or insufficient portrayal of a certain society's strata and groups.

Selection bias

Selection bias refers to instances where the data used to train an AI system does not represent the reality it intends to model. In other words, when the training data of a machine learning model is not representative of the environment in which the model operates due to how the data was chosen or gathered.

One prevalent type of selection bias is self-selection. This often arises in recommender systems, where users self-select themselves to express their preferences for certain items they choose to rate. Since the main objective of a recommender system is to suggest relevant content to users on a platform, these systems depend heavily on logs of content that users have previously shown a preference for [196]. Collaborative filtering strategies, which are integral to these systems, generate recommendations based on shared preferences among users. The underlying principle is that if a group of users shares similar tastes, items liked by some members of the group but not yet encountered by others can be recommended to the latter. As a result, users are more likely to be exposed to—and potentially rate—items that have already been favored by others. This also creates a feedback loop, where only a subset of items gathers preference data, while other items remain unrated, partly because they were never displayed to most users.

Similarly, content platforms, such as streaming services, represent users' preferences through explicit feedback, like ratings, or implicit feedback, such as time spent viewing content. Regardless of how user feedback is collected, the use of this data introduces a selection bias. This occurs because some content may gather most user preferences while others remain in the long tail of seldom-seen items. This type of selection bias is also known as popularity bias [185], where more popular items tend to be shared by more people and, therefore, are more frequently recommended. However, popularity is not always a proxy for content quality, as it is influenced by the item's visibility. Visibility can be manipulated by marketing campaigns for movies or songs and even by bots in the context of political campaigns on social media.

Search engines also exhibit this bias because they utilize user feedback to rank documents [110]. This results in ranking bias. Various studies have shown that documents at the top of a ranking are more likely to be selected by users, cementing their position over time, but not necessarily due to their true quality.

Another type of selection bias is non-random sampling, which occurs when data is collected in a way that does not provide each individual or data point in the target population with an equal probability of being included in the sample. When an AI model is trained on data derived from such a non-random or non-representative sample, biases may develop that disproportionately reflect the characteristics of the over-represented groups within the dataset. Consequently, the model's performance may be skewed, leading to inaccurate, unfair, or unreliable outcomes when it is applied to broader and more diverse populations.

For example, a credit scoring model trained primarily on data from wealthier populations may unfairly assess the creditworthiness of individuals from lower-income groups [189], thereby perpetuating economic inequalities. Similarly, in the criminal justice system, predictive policing models trained on biased crime data—often collected more heavily from minority neighbourhoods—can reinforce discriminatory practices, leading to over-policing of these communities.

Measurement bias

Measurement bias refers to the systematic error that occurs when the features or labels used in a predictive model do not accurately or consistently reflect the underlying constructs they are intended to measure. This bias arises when the data collected or computed as proxies for complex concepts fail to capture the full scope of the construct or when the measurement method (or its precision) varies across different groups.

Measurement bias emerge after constructing the sample [271]. This type of bias occurs when building the descriptors for the sample, specifically when selecting the characteristics and target variables of interest. These descriptive characteristics and the target variable serve as a proxy (a concrete measurement) to approximate an abstract (ideal) entity that cannot be directly encoded. Typically, what we measure is a one-dimensional reduction of a far more complex object. In this reduction, the proxy fails to capture the full complexity of the object it describes. For example, consider the challenge of predicting whether an employee will be effective in their job. The concept of "job effectiveness" is multifaceted and cannot be fully captured by a single measurable attribute. However, algorithm designers might use "years of experience" as a proxy for job effectiveness. This approach overlooks other critical factors, such as adaptability, communication skills, or creativity, which are important indicators of job performance but may vary significantly across different roles or industries. Consequently, relying solely on "years of experience" as a proxy can lead to biased predictions that do not accurately reflect the true potential of employees, particularly those from diverse backgrounds who may excel in other areas.

Measurement bias can also arise from disparities in the measurement of the proxy. For example, let us assume that in an educational context, we use grades as descriptors of student academic success. If the sample considers two generations, both grades are not directly comparable since each generation was assessed with different evaluation instruments. Consequently, drawing conclusions by comparing raw grade values across generations would not be appropriate because the scores do not mean exactly the same in different tests.

Furthermore, measurement bias may occur due to disparities in measurement effectiveness across different groups. For instance, pain scales used in healthcare are subjective measurements, with differences in pain perception across gender and racial or ethnic groups [315]. Since the pain perception scale is subjective, equivalent ratings for different individuals may represent different types of clinical complications, resulting in erroneous outcomes.

Measurement bias can also arise from the omission of critical variables. While using proxies to measure complex phenomena tends to oversimplify reality, excluding relevant variables disregards exogenous factors that were not considered during the construction of datasets descriptors. Consider the example of building a credit risk assessment system. After collecting data from banking clients, we can derive proxies to describe their financial behavior. The goal is to determine whether a client is creditworthy, with the system delivering a default risk rating—either high or low. Now, imagine that after the model has been trained, there are significant changes in the country's economic context, such as an adjustment in the monetary policy that leads to an increase in the benchmark interest rate set by the Central Bank. In this new scenario, the cost of credit rises, elevating the default risk. Consequently, the criteria for granting credit should become more stringent, leading to a higher proportion of individuals being classified as high-risk. However, because the model is not equipped to account for this change in economic conditions, it continues to classify some individuals as low-risk, even though they are now more likely to default. The group most affected by this measurement bias will be those whose creditworthiness is near the decision threshold, where the omission of these critical variables has the greatest impact on classification accuracy.

4.2.3 Algorithmic biases

Algorithmic biases refer to systematic and repeatable errors and can arise at various stages of the machine learning lifecycle, including model training and decision-making processes. Unlike random errors, algorithmic biases are embedded within the design and functioning of the algorithms, leading to consistent patterns of inequity in their outputs, which can make existing data biases more pronounced. Amongst these we can find biases introduced by the machine learning algorithm itself (inductive bias) [136], biases arising from how the model is constructed (aggregation bias) [216], or even decisions made during the experimental design phase that affect model selection (evaluation bias) [63]. These sources of bias are generally encompassed within the concept of algorithmic bias, which refers to any form of bias that is acquired or accentuated during the model training and model selection process. While all these sources are considered part of algorithmic bias, each has its unique characteristics.

Inductive bias

In machine learning, inductive bias refers to the set of explicit or implicit assumptions made by a learning algorithm to enable induction, which is the process of generalizing from a finite set of observations (training data) to a broader model of the domain [144]. Without such a bias, induction would be unfeasible, as the observations could be generalized in numerous ways. If all potential generalizations were treated equally, accurate predictions for new situations would not be possible without reflecting background knowledge about the target.

This happens because machine learning algorithms operate on the principle of identifying patterns and statistical regularities in data [136]. They tackle various tasks such as object recognition, classification, sequence labeling, and even content generation by processing datasets to recognize these patterns. These patterns can be represented in various ways, such as associations between variables, clustering patterns, descriptive or discriminative patterns, or object representations [48]. These representations may be designed by data scientists or learned during model training.

Regardless of how an algorithm represents detected patterns, its primary goal is to construct a function that extracts high-level descriptors from examples, such as a prototype in clustering algorithms or a target variable in predictive models. If the target variable is continuous, the resulting model is termed a regression model; if it is categorical, it is known as a classifier.

A key characteristic of machine learning is that, in constructing the function that maps data to the target variable, the algorithm prioritizes certain patterns while discarding others deemed less significant [116]. This learning process is guided by a function that evaluates the model's alignment with the data—either by directly assessing the model's performance on a task (for example, using an accuracy function on a categorical variable) or by evaluating the model's ability to preserve information from the original dataset with minimal loss. In some approaches, a beneficial relationship exists between task performance and information retention; minimizing information loss during model construction can directly enhance performance metrics for specific tasks.

Machine learning algorithms inherently prioritize certain patterns over others. While some patterns may help minimize information loss, others may be disregarded due to their lower relevance to the expected task outcomes. The decisions made to prioritize certain patterns constitute the inductive bias—the set of assumptions the model learns from examples that enable it to infer the target variable. However, prioritizing one goal over another can introduce challenges, such as maximizing overall task performance (e.g., classification accuracy), leading to poorer performance for underrepresented groups. Class imbalance, a significant challenge in machine learning, requires careful consideration due to its impact on model fairness and effectiveness.

Another factor contributing to bias during training is the pruning effect. Since models are compact representations of data, the emphasis on certain patterns over others can exacerbate disparities for underrepresented groups within the dataset. This occurs because these groups are smaller and, therefore, offer less descriptive richness. This effect is particularly pronounced in Large Language Models (LLM), a topic that will be explored further in subsequent chapters of this book.

Aggregation bias

This type of bias arises when data is combined or aggregated in a way that obscures important differences between subgroups within the data. For example, when models presume that the relationships or patterns identified at a higher level of aggregation (such as averages or general trends) are consistent across all subgroups, but this may not necessarily be the case. Different subgroups within the dataset may have distinct backgrounds, cultures, or norms, leading to variations in how certain variables should be interpreted [284], which can result in a model that is primarily fit for the dominant population or that is sub-optimal for any particular group, especially when combined with representation bias.

The assumption that a single model can capture the diversity of a dataset is questionable [216]. Often datasets amalgamate different populations, each exhibiting unique patterns. Forcing a single model to synthesize such a heterogeneous population for a task is overly ambitious. Like most conventional classifiers (e.g., logistic regression or support vector machines), traditional homogeneous models operate on this monolithic model assumption. This assumption leads to an aggregation bias, where the model, in its attempt to accommodate the dataset's diversity, may merge groups using the same descriptors, thus ignoring the nuances of each group. As a result, the model becomes sub-optimal for the groups within the dataset rather than accurately representing each group's distinct descriptors.

For instance, aggregation bias often occurs in datasets from social media [325]. Consider constructing a tweet classification model with the objective variable being the tweet's polarity, a typical task in social media analytics of sentiment analysis. Assume we collect tweets in California, USA, from both English-speaking and Spanish-speaking populations. While the use of emojis might differ in meaning between these groups, a monolithic model trained on this data would merge the examples into a single model, leading to aggregation bias. This bias ignores the specific context of each group, potentially leading to classification errors.

Aggregation bias relates to Simpson's paradox. According to it, conclusions drawn from analyzing a heterogeneous population as a whole may not hold when the population is broken down into strata. This is because observing aggregated trends across a heterogeneous population does not account for the particularities of each subgroup. For example, consider analyzing academic success on a standardized university entrance exam. Observations might show a bias towards the Caucasian population over the Indigenous population, with Caucasians entering more selective undergraduate programs while Indigenous applicants qualify for less selective courses. A possible conclusion might infer that the university selection system is biased against the Indigenous population, thereby perpetuating an injustice through a standardized test. However, upon disaggregating the data, we find that Indigenous applications typically target shorter, technical-professional careers, whereas most Caucasians apply to longer programs with higher selection thresholds. Simpson's paradox occurs in this scenario because the Indigenous population generally applies to programs with lower admission thresholds, not because the instrument restrictively channels them into such programs.

One way to mitigate aggregation bias is through non-monolithic models. Models that employ strategies to work with data partitions and fit specific models to each population segment are better at managing diversity, thereby reducing the effect of aggregation bias. Later in this book, we will explore various machine learning strategies that can lessen this effect, with ensemble models being particularly effective at handling data partitions.

Evaluation bias

Evaluation bias in machine learning refers to the systematic error that occurs when the metrics, methodologies, or datasets used to evaluate a model's performance do not accurately or fairly reflect its effectiveness across different contexts, groups, or tasks, thus potentially exacerbating bias in a model [279].

This type of bias can arise from the data or metrics chosen when training or evaluating a model. For instance, if a metric like classification accuracy is monitored during training, an under-representation of certain groups leading to class imbalance will cause accuracy to favor the performance of the majority classes. Consequently, the model will favor certain classes over others, resulting in disparate treatment. The disparity induced by the metric occurs because optimizing for accuracy overlooks the necessary balance between two fundamental measures: precision and recall. While accuracy is suitable in balanced contexts, classifiers tend to produce performance disparities between precision and recall in the presence of class imbalance. This disparity is related to an imbalance between the rates of true and false positives, as it probably happened in the COMPAS case (see details in Chapter 1). For example, to identify all targets in a class, we might make mistakes that improve recall, but if the dataset is imbalanced, this will increase the rate of false positives. This example shows that evaluation bias can stem from an inappropriate choice of the evaluation metric during model training.

This methodological deficiency can also arise after the model training phase, particularly during the model selection stage, when evaluating the model's performance on the testing partition. Evaluation bias at this stage can occur due to an improper choice of the testing data partition, where certain groups are either over-represented or underrepresented. Such imbalances can distort the evaluation process, leading to the selection of a model that is biased towards the over-represented groups within the testing data while inadequately assessing its performance on underrepresented groups. For example, suppose the testing partition predominantly consists of data from a specific demographic group, such as younger individuals in a healthcare application. In that case, the model may appear to perform exceptionally well during evaluation. However, this performance may not generalize to other demographic groups, such as older individuals, whose data was underrepresented in the testing partition. Consequently, the model selected for deployment may be suboptimal or even detrimental when applied to the broader, more diverse real-world population.

Furthermore, according to Suresh and Guttag [284], a model may be optimized on its training data, but its quality is often measured using established benchmarks such as UCI Machine Learning Repository datasets,[2] Faces in the Wild,[3] or ImageNet.[4] These benchmarks serve as a standard for comparing different models, allowing for quantitative evaluation. However, when these benchmarks do not adequately represent the full spectrum of real-world data, they can encourage developing and deploying models that perform well only on the subset of data included in the benchmark. This issue is more pervasive than other sources of bias because it operates on a broader scale. A misrepresentative benchmark can lead to the widespread adoption of models that appear effective based on evaluation metrics but fail in practice when applied to underrepresented groups or scenarios.

4.2.4 Other biases in machine learning

So far, we have explained in detail some of the most well-known biases in the process of generating a machine learning model. However, many more potential biases can emerge and affect the outcomes of a project that uses a machine learning model, as we illustrate in Figure 4.2. For example, previous research has suggested that historical biases influence the type of problems chosen for machine learning solutions, even before selecting the data, target variables and algorithms [81]. As a result, many current uses of AI can be linked, for example, to surveillance, while significantly fewer are aimed at addressing gender inequality or discrimination. This line of reasoning parallels discussions

[2]"UCI Machine Learning Repository," accessed August 19, 2024, https://archive.ics.uci.edu/datasets

[3]"Labeled Faces in the Wild (LFW) Dataset," accessed August 19, 2024, https://vis-www.cs.umass.edu/lfw/

[4]"ImageNet," accessed August 19, 2024, https://www.image-net.org/

about the objectives of scientific research, where, for instance, health issues that exclusively affect women have historically received less attention than other illnesses. Thus, developers and decision-makers are encouraged to reflect on how biases and societal influences shape the problems they aim to address with AI.

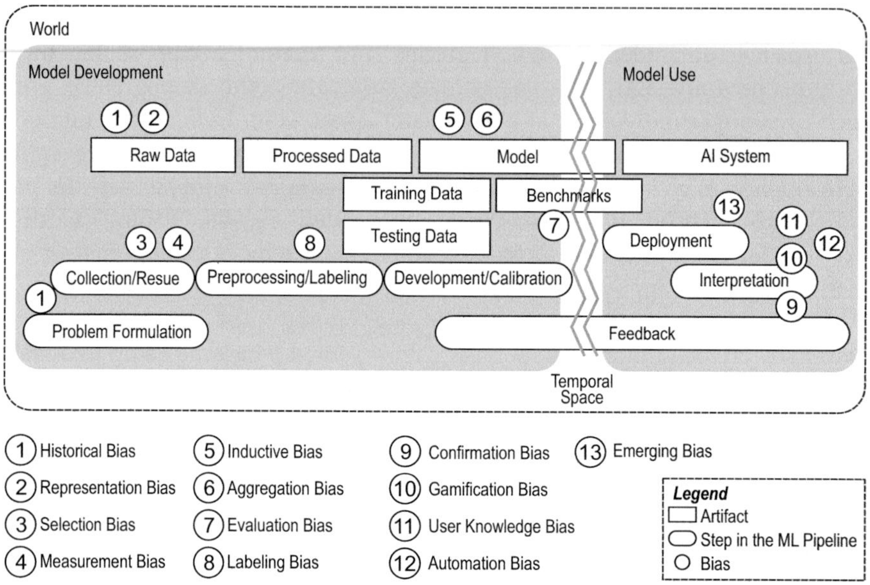

Figure 4.2: Examples of biases in the lifecycle of a machine learning project.

Other biases involve various stakeholders in the lifecycle of a machine learning project. For example, biases can occur when there is a need to label data or provide human feedback in reinforced learning because the people providing inputs can introduce their own biases in the metadata they are creating.

It is also critical to know that the creation of a machine learning model is not the end of a project. Such a model is often integrated into a system with a specific user interface, which is, in turn, embedded into an organizational process. Such a system and its user interaction is another scenario where biases can emerge. Users might be, for example, subject to automation bias, which leads them to rely on automated outcomes without critical examination. Rather than creating an exhaustive list of other potential biases along the machine learning lifecycle, we aim to communicate that there are more biases to be aware of; they are often related to each other and can influence the quality of a machine learning project. Therefore, it becomes necessary to consider ways to identify, monitor and control them over the course of an AI project. We recommend that this should be a collaborative effort where developers are actively engaged and can voice their concerns about biases. However, addressing these issues should be a shared responsibility, approached not only through technical means but also with an interdisciplinary strategy across various levels of the decision-making process.

4.2.5 Challenging the status quo: A Bias Network Approach (BNA)

In forthcoming work from Arriagada-Bruneau et al. [329], we propose challenging the prevailing methods identifying biases in AI projects, by proposing a sociotechnical solution. The Bias Network Approach (BNA) is a method designed to address biases in AI development by mapping and visualising their interconnections, rather than treating them as isolated occurrences. This proposal seeks to counter what we term the "isolationist approach," which narrowly focuses on individual biases at specific stages of an AI pipeline (as illustrated in Figures 4.1 and 4.2). By contrast, the BNA aims to foster ethical reflection through guided dialogue among developers, facilitated by interdisciplinary experts, to elucidate the connections between biases, their sources, and their impacts on the outcomes of an AI project. We piloted the BNA with a healthcare-focused natural language processing (NLP) project in Chile. The results demonstrate the BNA's effectiveness in fostering transparency, highlighting interconnected biases, and encouraging developers to consider broader societal and professional influences in their decision-making. Crucially, the BNA identifies material limitations, external factors, and professional biases as significant sources of bias, which are often overlooked in traditional AI bias literature. For example, material limitations—such as resource constraints and inconsistent data quality—played a pivotal role in shaping decisions and introducing biases throughout the healthcare project. Similarly, professional biases, stemming from developers' engineering-oriented training, led to an overemphasis on technical performance metrics (e.g., F1 scores) at the expense of broader ethical considerations, underscoring the need to address such biases more explicitly within AI ethics frameworks. The BNA's visual network mapping emerged as a critical tool for ethical reflection and transparency. By illustrating how biases interconnect across development stages, rather than appearing as discrete instances, developers gained a more comprehensive understanding of the ethical implications of their decisions. This mapping facilitated collaborative discussions within the team, enabling the articulation and mitigation of previously unrecognised biases. Furthermore, developers noted the practical utility of the visualisation for improving transparency in internal processes and for external communications with stakeholders such as government agencies. The BNA aims to transcend passive compliance with ethical checklists by embedding ethical reflection as a dynamic and integral part of the AI development process. Developers acknowledged the method's potential for application at various stages of AI projects, from experimental design to retrospective evaluation. This adaptability positions the BNA as a valuable framework for enhancing ethical awareness and decision-making throughout the AI lifecycle. Additionally, this sociotechnical method aligns very closely with the concept of moral imagination, i.e., the ability to identify and critically reflect on ethical aspects of decision-making that might not be immediately apparent, as well as to creatively envision alternative perspectives and solutions. This concept, as discussed in the work of Lange et al. [330], is key to address complex ethical dilemmas, particularly in sociotechnical systems like AI development because it can promote:

(i) recognising the limitations of one's perspective as AI developers must acknowledge that their understanding of a situation, including available options and the ethical factors at play, might be incomplete or biased. This requires moving beyond the default, narrow focus often shaped by professional or disciplinary constraints (e.g., prioritizing technical performance metrics like accuracy or F1 scores over societal implications), and

(ii) creatively exploring alternative perspectives because developers are encouraged to imagine new approaches or solutions that consider overlooked ethical considerations, diverse viewpoints, and potential long-term impacts of their decisions. The BNA, through its participatory and reflective nature, operationalises moral imagination by promoting critical and creative thinking about biases in AI. By encouraging developers to see AI as part of a complex web of human, societal, and technical interactions, the approach aligns with the concept's aim of enabling anticipatory governance and ethically grounded decision-making.

We encourage readers interested in the BNA to explore the interconnectedness of biases and their broader sociotechnical implications, which can support in:

- Moving beyond a "microscopic" focus on technical aspects and consider the broader context, including societal and professional biases.
- The anticipation of ethical challenges and trade-offs at various stages of AI development, fostering a more comprehensive ethical awareness and preventing further complications.
- Questioning assumptions and exploring alternative paths, improving the ability to foresee potential harms or unintended consequences.

4.3 Further readings

An example of biased outcomes in a machine learning system and an analysis of how alternative technical decisions could have mitigated such biases is discussed in [222]. Surveys on biases and their connection to various stages of data-driven projects can be found in [284, 70]. The work of Friedler et al. [106] explains the gaps between the concepts people intend to represent, what is actually measured, and its implications for fairness in AI. For further insights on disparate impact, we recommend the work of Barocas and Selbst [22].

Ricardo Baeza-Yates [16] addresses biases on the web and highlights the need for more transparent algorithms to avoid perpetuating these biases. For discussing biases in social data, we recommend Olteanu et al.'s work [225]. Larsson and Heintz [180] discuss the importance of transparency in AI systems, arguing that it is crucial for accountability and trust. They emphasize that, without transparency, it becomes challenging to diagnose and rectify biases in AI systems. These papers collectively underscore the importance of addressing and mitigating bias in AI. They highlight the necessity of developing robust, fair, and accountable algorithms to foster trust and equity in deploying AI systems within society.

Meredith Broussard's book "Artificial Unintelligence" discusses use cases where not only do threats materialize, but also tangible harms arise from the misuse of these technologies [42]. Broussard highlights that the biases inherent in these machines are not merely replicated but exacerbated, leading to scenarios where humans, by over-relying on AI, lose sight of its limitations. Similarly, Broussard demonstrates the pernicious effects of AI biases. In "More than a Glitch" [43], she presents case studies that reveal biases related to race, gender, and access to technology. Relatedly, we report in a study of misinformation in Chile that the gap in technology access and understanding of technology defines a particularly vulnerable audience [198]. This vulnerability makes them more susceptible to the negative impacts of AI-driven misinformation. Along the same lines, Charlton McIlwain, in "Black Software" [193], shows how racial biases are exacerbated on digital platforms. He illustrates how intelligent systems trained on web data perpetuate racial stereotypes, further reinforcing the relationship between these threats. This exacerbation of biases through AI systems trained on data reflecting societal inequalities underscores the need for a critical approach to developing and deploying these technologies. It highlights the importance of ensuring that AI systems are designed and implemented in ways that are aware of and actively counteract these inherent biases.

Chapter 5

Fairness, Accountability, and Transparency in AI

Fairness, accountability, and transparency are essential principles in AI as sociotechnical systems [21]. These principles help characterize the ecosystem in which an AI is developed, taking into account not only the design and development stages but also its entire lifecycle [285]. This chapter addresses these three concepts, which are fundamental to understanding the ethical considerations surrounding AI, its use, and what will later be recognized as regulatory efforts.

5.1 Fairness in AI

We will begin by addressing the concept of fairness in AI. There are various definitions of fairness in the literature, and no single conceptual definition is considered predominant in the field [245], but this is not unique to AI contexts.

Fairness, as a philosophical concept, has long been a subject of deep inquiry, eluding a definitive and universally accepted definition. The difficulty in defining fairness stems from its inherently complex nature, entwined with various moral, social, and political dimensions. Philosophers have debated the essence of fairness for centuries, considering it not only in terms of equal treatment or justice but also through lenses of equity, entitlement, and social justice. This complexity is mirrored in the myriad of philosophical approaches that attempt to encapsulate what it means to be fair—each offering a perspective that is often context-dependent and influenced by cultural, societal, and temporal factors.

For example, fairness is often aligned with distributive justice, concerned with the equitable allocation of resources or benefits among individuals, and associated with John Rawls' seminal work A Theory of Justice [246]. For him, the key concept is based on a "justice as fairness" framework, proposing the difference principle, through which —put into simple terms—he claims fairness is achieved when inequalities are arranged to benefit the least advantaged members of society.

Beyond distributive justice, fairness extends to procedural justice [201], which considers the fairness of the processes that lead to outcomes. Procedural fairness is typically assessed without considering the outcomes it generates, emphasizing instead on whether the processes follow impartial and consistent rules. This dimension introduces further complexity, as fairness must be assessed not only in terms of outcomes but also in the mechanisms by which those outcomes are achieved. For instance, a procedure that consistently uses biased criteria may appear to produce "fair" outcomes when evaluated only on consistency. However, it fails to meet broader ethical standards of fairness, a challenge evident in how we address biases using fairness metrics in AI.

The complexity of fairness becomes even more pronounced when we consider interactional fairness [69], which pertains to the treatment of individuals within processes or systems. This concept underscores the relational and contextual dimensions of fairness, acknowledging that the way individuals are treated in interpersonal interactions can greatly shape their perceptions of fairness, regardless of procedural or distributive factors. This is often studied in work settings and organizational contexts. However, it can also be considered when studying fairness in recruitment practices that rely on AI models trained in biased historical data.

The ethical challenges of AI fairness arise precisely because AI systems are not neutral; they are embedded in, and often reinforce, existing social inequalities and biases. This embedding necessitates a shift from a narrow, metric-driven approach to fairness towards a more holistic ethical framework that considers the broader impact of AI on society.

Developers and other stakeholders in the AI ecosystem, must be aware of the risk of reducing fairness to a set of quantitative metrics. While metrics such as demographic parity or equal opportunity are important tools for identifying bias in AI systems, they are inherently limited. These metrics often fail to capture the full ethical landscape of fairness, particularly when they are applied without considering the broader context in which AI systems operate. This, however, remains a human job, and it is our responsibility to actively engage with this broader framework to think about fairness in AI.

Thus, the ethical implications of AI fairness extend to the processes and decisions that shape AI systems. Fairness must be considered at every stage of the AI lifecycle, from problem formulation, data collection and model development to deployment, use, and feedback instances. This process-oriented approach to fairness recognizes that ethical considerations cannot be an afterthought but must be integrated into the design and implementation of AI systems from the outset.

Hence, fairness in AI necessitates a framework that does not merely focus on metrics or outcomes but scrutinizes the processes that lead to those outcomes, the social contexts they affect, and the power structures they interact with. AI systems are often created by and for those in positions of power, which can lead to the reinforcement of existing social hierarchies. As argued by Virginia Eubanks [97], AI systems, when deployed in social welfare contexts, can perpetuate structural inequalities even when fairness metrics are considered. AI fairness, therefore, cannot be disentangled from the socioeconomic systems that these technologies interact with. Pursuing an ethical and fair development of AI systems requires an awareness of these power dynamics and a commitment to ensuring that AI systems serve the interests of all, particularly marginalized and vulnerable populations.

To date, fairness has been operationalized in different ways in AI. These operationalizations of fairness can be categorized into two groups depending on whether they assess fairness at the level of groups or individuals [318]. These definitions are based on the principle of fair treatment, meaning they aim to produce similar outcomes for similar individuals or groups [302]. This implies the definitions focus on the outcomes generated by AI models and on detecting disparities for individuals or groups in relation to these outcomes [49].

5.1.1 Group-based definitions

An AI model satisfies the concept of group fairness if different demographic groups within a population have an equal probability of being classified into a specific category [195]. This definition incorporates a set of demographic variables for analysis, which include commonly used categories such as gender, sexual orientation, religion, race, ethnicity, and disability [183]. Generally, we focus on gender in our definitions, but it is important to note that other demographic variables are also considered in the context of fairness in AI [8].

Consider a credit risk model where the target variable Y is a binary variable indicating whether an individual is creditworthy. The model is deemed fair with respect to gender if $P(\hat{Y} = 1|G = m) = P(\hat{Y} = 1|G = f)$, where $\hat{Y}$ is the model's prediction and G is the group descriptor, in this case, gender. The primary idea is that applicants have an equal chance of obtaining credit regardless of their gender.

Conditional statistical parity extends this notion by including a set of attributes expected to legitimately affect the outcome. For instance, in credit assignment, legitimate predictive attributes might include the amount of credit requested, the applicant's age, employment, and financial history. These factors are considered control variables in the analysis. Thus, under similar conditions regarding these control variables, the classifier should not introduce disparities in the outcomes for the two groups [134]. Let L be the set of legitimate attributes used as control variables. The classifier maintains conditional statistical parity for gender if $P(\hat{Y}=1|L=l,G=m)=P(\hat{Y}=1|L=l,G=f)$.

These definitions extend to the evaluation phase of the classifier. A key idea in this extension is to ensure that the rate of correct predictions is equal across different groups. This leads to the notion of predictive parity, meaning $P(Y=1|\hat{Y}=1,G=m)=P(Y=1|\hat{Y}=1,G=f)$. If the classifier is binary, as in credit assignment, predictive parity also implies $P(Y=0|\hat{Y}=1,G=m)=P(Y=0|\hat{Y}=1,G=f)$. The definition includes a balance in false positives, known as false positive error rate balance, defined as $P(\hat{Y}=1|Y=0,G=m)=P(\hat{Y}=1|Y=0,G=f)$, and false negative error rate balance, defined as $P(\hat{Y}=0|Y=1,G=m)=P(\hat{Y}=0|Y=1,G=f)$. Mathematically, a false negative error rate implies $P(\hat{Y}=1|Y=1,G=m)=P(\hat{Y}=1|Y=1,G=f)$, a condition known as equal opportunity. This means that the probability of a person with a good credit history obtaining a loan should be equal for both groups.

Another fairness notion is called equalized odds, also known as conditional procedure accuracy equality or disparate mistreatment [195, 319]. Equalized odds combine two conditions: a classifier meets the fairness definition according to equalized odds if the groups being analyzed have equal true positive and false positive rates. This is equivalent to the conjunction of the definitions of false positive error rate balance and false negative error rate balance, i.e., $P(\hat{Y}=1|Y=i,G=m)=P(\hat{Y}=1|Y=i,G=f)$, where $i\in\{0,1\}$. The idea behind this definition is that both good and bad credit applicants should perform comparably in the classifier, regardless of the analyzed group [134]. Similarly, we can measure conditional use accuracy equality as the conjunction of predictive parity fairness conditions, i.e., $P(Y=1|\hat{Y}=1,G=m)=P(Y=1|\hat{Y}=1,G=f)$ AND $P(Y=0|\hat{Y}=0,G=m)=P(Y=0|\hat{Y}=0,G=f)$. This definition implies equivalent accuracy for both groups in both classes. In the example of credit assignment, the definition implies that the probability of an applicant with good credentials obtaining credit is equal to the probability of an applicant with poor credentials not obtaining credit.

An additional definition of fairness in AI focuses on measuring the errors made by the classifier rather than its accuracy [195]. Treatment equality is based on the ratio of false positive and false negative errors. A classifier is considered to have treatment equality between two groups if their error ratios are equal, i.e., $\frac{FN}{FP}|\text{male}=\frac{FN}{FP}|\text{female}$.

Furthermore, another group-based definition of fairness considers a classifier's predicted probability score s, instead of the predicted label $\hat{Y}$ [162]. Test fairness, also known as calibration or matching conditional frequencies, is defined as predictive parity, which considers the fraction of positive class predictions for any given value of s. In the context of credit allocation, test fairness implies that $P(Y = 1|S = s, G = m) = P(Y = 1|S = s, G = f)$. Classifiers usually only partially meet the test fairness criterion because the classifier's performance tends to decrease for values of s close to the decision boundary threshold—i.e., $s \to 0.5$. This occurs because examples near the 'confusion zone' between classes tend to accumulate a higher number of misclassified examples.

5.1.2 Individual-based definitions

In the context of defining legitimate attributes used in group-based definitions such as conditional statistical parity, individual-based definitions rely on either awareness or unawareness of sensitive attributes [163]. These definitions are premised on identifying a set of sensitive attributes that the classifier should not consider when predicting the target variable. A key consequence of this approach is that two individuals with identical non-sensitive attributes should receive the same outcome from the classifier [183]. This concept leads to what we term "fairness through unawareness." A classifier is considered unaware if it does not use sensitive attributes to generate its outcomes. For example, in credit assignment, a classifier is unaware of the sensitive attribute of gender if this attribute is not considered during the training process. Fairness through unawareness implies that two subjects, i and j, who have the same non-sensitive attributes should receive the same result; that is, if $X_i = X_j$, then $\hat{Y}_i = \hat{Y}_j$. This outcome also suggests that the dataset does not use a gender proxy to produce the result.

Another definition of fairness based on individuals is called fairness through awareness [89], which captures the principle that similar individuals should receive identical outcomes. In this context, the similarity of individuals is measured by a distance metric. If the classifier is fair, the distance between the output distributions of individuals should be no greater than the distance between their representations. Formally, for a set of individuals U, an individual distance metric $d : U \times U \to \mathbb{R}$, a mapping from U to a probability distribution over model outcomes $M : U \to P$, and a distributional distance metric $D : P \times P \to \mathbb{R}$, we say that the classifier is fair through awareness for two individuals i, j if and only if $D(M(i), M(j)) \leq d(i, j)$. For example, consider two subjects i, j whose representations X_i and X_j have a Euclidean distance of 0.3. The credit risk model yields the following probabilities: $P(\hat{Y}_i = 1|X_i) = 0.6$ and $P(\hat{Y}_j = 1|X_j) = 0.7$. Distributionally, given the binary nature of the classifier, the subjects receive the probabilities $M(i) = [0.6, 0.4]$ and $M(j) = [0.7, 0.3]$. The typical probability metric used is the statistical distance, also known as total variation, denoted as $D_{\text{tv}}(P, Q) = \frac{1}{2} \sum |P(a) - Q(a)|$, which corresponds to half of the L1 norm (absolute difference). In our case, $D_{\text{tv}} = 0.1$. Since $D_{\text{tv}}(M(i), M(j)) = 0.1 \leq d(X_i, X_j) = 0.3$, we state that the classifier is fair through awareness for individuals i and j.

5.2 Causal reasoning

Group-based and individual-based definitions operate at different levels. While group-based definitions highlight the importance of demographic characteristics and, based on these, define fairness conditions among groups, individual-based definitions focus on the use or non-use of protected attributes, such as an individual's membership in or identification with a group. Both group-based and individual-based definitions rely on the use of sensitive attributes during model construction or on the possibility of identifying differences in model outputs based on these attributes.

An alternative approach involves assessing whether these attributes actually have an effect on the model's output. This approach leads to a causality analysis, which entails exploring whether there is indeed evidence to support that a specific attribute causes a particular output.

Definitions of fairness based on causal reasoning are founded on the principle that a decision is fair for an individual if the decision would remain the same under the current conditions or under alternative conditions that result in changes to protected attributes [169]. For instance, a classifier is considered fair under causal reasoning with respect to the protected attribute of gender for an individual if the classifier's outcome would be the same regardless of gender [47].

In causal reasoning, the dependency between variables is formalized using directed graphs. The concept of a directed dependency graph involves establishing causal dependency relationships between variables. Causality is not the same as correlation. In a causal relationship, the link is directed from a precedent (cause) to a consequent (effect).

A common way to analyze a model from the perspective of causal reasoning is to define causal graphs between variables. A causal graph is a directed acyclic graph where nodes represent variables and edges represent relationships between these variables. Causal graphs are used to analyze classifiers.

Causal graphs consider different types of nodes. A node is a proxy if the value of the variable it represents can be used to determine the value of another variable in the graph. This type of relationship between a proxy variable and a derivable variable is represented in the causal graph with a directed edge from the derivable node to the proxy node. For example, in the case of credit assignment, let us assume we have a Boolean variable called 'Turner Syndrome,' which indicates whether the person applying for credit has this syndrome. Turner Syndrome is a genetic disorder that affects the sexual and reproductive development of girls. Since it can only affect women, the 'Turner Syndrome' variable is a proxy for gender. A causal graph would represent this relationship with a directed edge from 'Gender' to 'Turner Syndrome'.

The causal relationship between a variable and its proxy must be carefully considered if the independent variable is protected. For example, if gender is a protected variable, using 'Turner Syndrome' as a proxy—even though the gender variable itself is not used in the classifier—will result in gender-biased outcomes, thus making the model unfair with respect to the protected variable.

If the relationship between a protected variable and a dependent variable does not alter the outcomes of the classifier, the dependent variable is termed a resolving attribute. Let's consider the variable 'Credit Amount', which is dependent on the protected attribute 'Gender'. This dependency arises because, on average, women apply for smaller loans than men. However, if we assume that the 'Credit Amount' does not influence the classifier's decisions—meaning credits are granted for both low and high amounts—then 'Credit Amount' would be considered a resolving attribute for 'Gender'.

A causal graph can include different types of relationships between variables. In the context of analyzing fairness, these relationships are of interest only if they involve protected attributes or proxy attributes.

From the analysis of causal graphs, we have two approaches to fairness in machine learning (ML) based on causal reasoning. We say that a classifier is counterfactually fair if its outcomes do not depend on a protected attribute [177]. For example, a classifier is not counterfactually fair if it utilizes the 'Turner Syndrome' variable. Although the classifier does not utilize the protected attribute 'Gender' (thus achieving fairness through unawareness), it fails to be counterfactually fair because it employs 'Turner Syndrome', which is a proxy for 'Gender'. We define a classifier as free from proxy discrimination if there is no path from a protected attribute to the classifier's output that is mediated by a proxy.

It is possible for a classifier to be free from proxy discrimination but not be counterfactually fair. This situation occurs if the classifier does not use the variable 'Turner Syndrome' but does use 'Gender'. In this case, there is no proxy for the protected attribute, but the classifier is unfair because it uses the protected variable 'Gender'.

It is important to note that counterfactually fair and proxy discrimination are two complementary definitions of fairness in ML. However, the example shows that proxy discrimination is a weaker definition of fairness than counterfactually fair. This is because a model can satisfy the definition of proxy discrimination but fails to meet the definition of counterfactually fair.

Overall, there are many ways to measure and assess fairness in AI, but the challenge lies in determining which mechanism best aligns with the specific conceptualization of fairness needed for a given project. Previous research has suggested that it is mathematically impossible to satisfy different definitions of computationally-measured fairness metrics simultaneously [172]. Consequently, AI fairness frameworks have proposed guidelines on which metrics to prioritize based on the types of predictions made by each particular AI [74]. Nevertheless, this area of research and practice is actively evolving and we expect to see new conceptual and technical developments in the coming years, especially moving beyond predictive AI and addressing the challenges of generative AI.

5.3 Accountability in AI

While transparency is crucial for understanding how AI systems make decisions, the concept of accountability brings us to think about how these systems are held responsible for their outcomes and actions. The development of intelligent systems offers opportunities to enhance the efficiency of various activities but also introduces new scenarios involving unexpected consequences and fundamental ethical challenges.

Accountability is crucial because intelligent systems allow us to delegate tasks to algorithms. Although there are many definitions of accountability, especially in documents related to AI governance like the GDPR [236] or the ALTAI [139], they all agree on one aspect: accountability is the obligation to report and justify actions to an authority. Following Bovens [39], accountability is an umbrella concept that encompasses and overlaps various notions such as transparency and responsibility. At its core, accountability involves a relationship between an actor and a forum, where the actor must explain and justify their conduct, the forum can pose questions and assess, and as a result, the actor may face consequences.

The actor can be an individual, but it can also be an organization (public or private). On the other hand, the forum might be an individual endowed with authority, typically with a public connotation, such as a journalist, a judge, or a prosecutor. It is important to note that public connotation does not necessarily relate to government. The public authority might be significant because the individual fulfils a public role, such as informing. In this scenario, whether the individual belongs to the government or a private entity is irrelevant. The individual is vested with authority by public connotation due to their role. The forum could be a state entity, like a judicial court, or a governmental body, such as a regulatory or supervisory agency.

The duty of the actor to report to the forum can arise in both legal and informational contexts, which are not mutually exclusive. An informational instance aims to gather sensitive information based on the actor's testimony, while legal responsibilities stemming from this information might later fall under a legal instance that could involve potential sanctions.

Accountability takes various forms depending on the actors and the forum involved. There is political accountability, which elected officials are subject to. There is also legal accountability, which applies to citizens within a rule-of-law framework and private legal organizations. In both political and legal accountability, forums may impose formal sanctions as defined within the legal framework of each country. Additionally, there is administrative accountability, where forums consist of auditors and regulators acting independently. This type applies to financial supervision, external process controls, and quality accreditation agencies. The actors in this type of relationship are generally public or private organizations. There is also professional accountability, where the actors are professionals, and the forums are professional associations and guilds that oversee the ethical practice of their profession. Finally, there is social accountability triggered by growing mistrust in state institutions and their regulatory roles. In the absence of independent regulatory forums, non-governmental organizations often launch self-managed initiatives to fulfill this role. Since these instances lack sanctioning powers within the legal framework, they do not impose legal or administrative sanctions but moral ones. However, these types of forums might later turn to the judiciary if there are suspicions of legal responsibility.

In the context of AI, several types of accountability relationships are involved. The clearest is legal accountability, as designers, developers, and corporate officers who deploy and popularize intelligent systems must comply with the defined legal and regulatory framework. If the regulatory framework regarding this aspect is lenient, social accountability comes into play, meaning that civil society should perform oversight. These quasi-legal instances can disclose information and establish social sanctions. There may also be political and professional accountability when AI, such as generative AI, is used for propaganda.

Furthermore, a key obstacle for AI accountability is the lack of transparency that surrounds AI systems. Opaque AI models obscure the factors contributing to their unfair outcomes or other forms of harm. For example, these outcomes may depend on the data with which the AI model was trained but could also be influenced by the learning algorithm, which may exacerbate or distort certain patterns. Since deploying AI systems in organizational settings inherently delegate tasks to AI, in terms of accountability, responsibility becomes diluted among the various factors that could lead to undesirable outcomes. Tracing the causes of such results in an opaque system remains a challenging task.

5.3.1 What characterizes accountability in AI?

According to Novelli et al., [220], the main characteristics of accountability in AI include context, scope, and agents involved. Context refers to the field in which AI is being used and the AI system's autonomy level. Scope involves the specific stages of a process where AI is applied, such as design, development, or deployment. Design involves tasks such as planning, architectural design, selecting foundational technologies or models, interface design, data usage, and the development strategy employed. Development relates to the composition of teams, which could include programmers, engineers, testers, or team coordinators. Deployment involves monitoring whether the system delivers expected outcomes and maintaining these AI systems. The third characteristic, agents, can be identified individually, corporately, collectively, or hierarchically.

This configuration is crucial for understanding the type of AI accountability in question. AI accountability can be either reactive or proactive. Proactive accountability is seen as a process virtue, integrated within the planning purposes to anticipate events and prevent failures. Conversely, reactive accountability occurs after harmful events have taken place, aiming to mitigate failures that have already occurred.

AI accountability mechanisms can involve recommendations, approvals, prohibitions, or various types of sanctions. Various nations and organizations are working on developing accountability standards, with many aiming for a global scope. These efforts involve countries adhering to international treaties or multilateral agreements that cover diverse areas, including technology applications and commercial standards. The application of these standards necessitates defined processes like internal audits, self-assessments, peer assessments, or external evaluations, all linked to monitoring the effectiveness of systems and their compliance with standards.

In this context, the role of nation-states is crucial. Nation-states, through their various governmental organizations, are responsible for regulating this domain by developing frameworks that encourage AI development while safeguarding civil society from associated risks and potential adverse scenarios. These efforts are referred to as AI governance [287].

Despite the fundamental importance of AI governance for responsible AI development, progress in this area is still in its infancy. There are no clear definitions of what AI governance entails or its scope. The core of divergent views on AI governance hinges on our perception of the role of the States. While some perspectives advocate for minimizing the State's role in technological development, leaving the field open to large transnational IT companies, others call for the development of public policies within clearly defined regulatory frameworks. Issues such as data privacy, the ethical implications of using AI in decision-making, or workforce replacement are key areas of further discussion in this realm.

5.4 Transparency in AI

In addition to fairness, transparency is another crucial aspect of AI ethics. AI, particularly deep learning, faces scrutiny for its lack of transparency regarding how it makes decisions or generates content [51, 303]. This often leads to it being described as a "black box." The transparency issues exacerbate the concerns about unfair outcomes, as the opacity of AI models makes it more challenging to detect and address these problems. Beyond fairness, transparency issues also encompass other ethical considerations. For instance, researchers have raised concerns about the extensive use of private data without clear and accessible information on its intended use [258] and the considerable environmental impact associated with AI technologies [67, 188].

Conceptually, transparency is linked to the idea of understanding. According to Michael Reddy [247], understanding depends on a systematic set of operations that move from the domain of physical objects to the domain of mental operations. Understanding is deeply connected with the concepts of seeing and knowing, both of which relate to comprehension. Furthermore, terms such as illuminate, clarify, and make transparent are linked with the act of understanding. What is not transparent and thus obscure or opaque is considered incomprehensible. Thus, a key goal of AI transparency is making AI understandable.

In the realm of AI, transparency is operationalized in several ways [180]. One perspective relates transparency with the concept of openness. A transparent system can be an open system. An intelligent system is deemed open if it involves open data, open source, and open access.

The openness of a system is determined by the business model of the corporation that developed it or the principles that inspire a community of developers. For instance, communities that advocate for open data or open source encourage practices that foster the design and development of open intelligent systems. Openness fosters conditions conducive to reproducible results. Thus, reproducibility and openness are interconnected.

Conversely, proprietary systems, which use proprietary licenses, tend to be closed systems. It is more challenging to discern the rules, patterns, and models used by such AI systems to produce results. While not all proprietary systems are completely closed, they typically exhibit limited openness. For instance, proprietary AI systems might reveal the data they were trained on, but accessing their code is typically more challenging. Open and closed systems can coexist on the same platform. For instance, HuggingFace, a well-known platform for reproducibility, supports both types of initiatives. In this context, a closed system can possess features that enhance reproducibility [116]. However, the link between an open-source system and reproducibility is evidently stronger. On HuggingFace, closed systems provide executables in such a way that they act as black boxes, enabling us to operate these models on the platform and facilitating the reproducibility of results.

Another crucial perspective related to transparency is explainability. A system's explainability represents the ability to extract explanations from an AI model. Explainability leads to what is termed Explainable Artificial Intelligence (XAI), which consists of a set of methods that enable black-box models to produce explanations that enhance transparency. We will revisit the mechanisms to generate explanations from AI models later in the book; however, we will discuss here the role of explainability in achieving transparency.

Unlike openness, this perspective aims to tackle the opacity of AI models that hinder understanding the processes driving a given output. This issue is especially problematic for systems using deep learning [13]. These models often function as black boxes, enabling developers to verify if they produce the expected outcomes for specific inputs (using test data) but not to elucidate the internal mechanisms behind these results. As a result, even developers are unable to fully grasp and communicate the reasoning behind these AI decisions to users. This lack of transparency prevents users from assessing whether an AI decision was reached through a rational or appropriate method, and it obstructs their ability to make a well-founded appeal against the decision or to devise a strategy for securing a more favorable outcome [296].

XAI mechanisms aim to offer either global explanations that apply to all outcomes of an AI model or local explanations tailored to specific outcomes. However, XAI is mainly designed by and for developers, frequently neglecting the perspectives of end-users and those affected by AI [207]. There remains a need for standard, effective methods to inform non-technical users about AI and its functioning [115, 282]. A younger field called human-centred explainable AI (HCXAI) has shifted the emphasis from the technical aspects of generating information about AI models to the challenges of effectively communicating this information to individuals who need to understand and use it for decision-making [91]. HCXAI is consistent with the concept of "meaningful transparency," which seeks to ensure that individuals understand AI decisions in a way that is relevant to them. This understanding would allow them to intervene, approve or reject decisions, and hold the AI accountable [40]. The literature identifies several groups that need explanations about AI models, including AI developers, domain experts (such as judges receiving predicted recidivism scores), affected individuals (such as the accused), general users, AI auditors, and policy-makers [252].

So far, HCXAI research has focused on evaluating whether people understand explanations, find them usable, develop appropriately calibrated levels of trust in AI, and whether explanations improve AI-human collaboration. Evidence indicates that while AI explanations often improve users' subjective understanding, their impact on objective understanding and trust is mixed [256]. Some explanations have failed to foster AI acceptance [156] and have sometimes resulted in unwarranted trust [324] and overreliance [20]. Thus, and somewhat unexpectedly, researchers have found other risks associated with AI explainability. Ehsan and Riedl [92] discuss "explainability pitfalls," where AI explanations might lead users to overly rely on AI decisions at the expense of their own judgment. HCXAI is making progress, and we anticipate increased research in this area that will complement advancements in XAI. However, a major gap that needs to be addressed is that most research has concentrated on contexts within the Global North, leaving a lack of understanding about how XAI techniques are developed, implemented, or assessed in communities in the Global South [224].

Lastly, there are various other views on transparency. Some approaches aim to enhance the transparency of decisions made during AI training and evaluation processes. For instance, certain mechanisms focus on improving traceability regarding data and model choices, such as documenting decisions in datasheets for datasets [113] and model cards [208]. Other methods involve conducting algorithmic audits [200] to evaluate the fairness of AI models and to scrutinize the processes that lead to AI outcomes. Significant effort is still required to implement transparency in AI projects, addressing aspects such as processes, users, models, and data. A promising approach to distinguish between these aspects of transparency is offered by the conceptualization of transparency by design, which identifies three levels of transparency: transparency by virtue, relational, and systemic transparency (Felzmann et al., 2020). Transparency by virtue refers to the act of disclosing information about the internal operations of an AI system, where we can position most XAI research and other efforts to make public model documentation and evaluation. Relational transparency focuses on how people perceive and understand this information, highlighting that there are different kinds of transparency users and efforts need to be done to assess how their information needs are being satisfied. Thus, HCXAI fits in this second level. Finally, systemic transparency involves the institutional context where these connection between an AI system and its transparency stakeholders.

We expect substantial advancements in these areas in the future as current regulations and recommendations increasingly focus on these directions.

5.5 Further readings

Algorithmic fairness and bias are addressed in several papers, including those by Friedler et al. [107], Corbett-Davies et al. [63], and Feldman et al. [99], that explore the pervasive issue of bias in AI algorithms. Friedler et al. [107] evaluate interventions for enhancing fairness in machine learning and emphasize the trade-offs between different types of fairness. Corbett-Davies et al. [63] explore the cost of fairness in algorithmic decision-making, demonstrating that efforts to make algorithms fairer can sometimes lead to other unintended inequalities. Feldman et al. [99] discuss methods for certifying and mitigating disparate impact, which is crucial for developing AI systems that do not systematically disadvantage any particular group. The Aequitas' audit framework [74] and IBM Fairness 360 [249] tool allow the computation of several fairness measures.

Regarding debiasing techniques and counterfactual fairness, the challenge of debiasing AI is thoroughly addressed by Bolukbasi et al. [35] and Kusner et al. [177]. Bolukbasi et al. present methodologies for debiasing word embeddings [35], critical for reducing gender and ethnic stereotypes in NLP applications [47]. Kusner et al. [177] introduce the concept of counterfactual fairness, proposing a model that ensures fairness of decisions across different demographic groups under a counterfactual scenario where sensitive attributes might vary. This approach helps understand and reduce biases ingrained in predictive modelling.

Regarding accountability, Raja et al. [331] provide an extensive survey of AI accountability, emphasizing its operationalization across the AI lifecycle. Their analysis identifies critical factors influencing accountability at various levels, including data, algorithms, and developers. The authors propose mechanisms such as decision provenance and algorithmic impact assessments as essential tools for fostering responsible decision-making and mitigating risks .

Building on this, Birhane et al. [332] critically examine the role of AI audits as a pivotal mechanism for achieving accountability. Their work identifies significant gaps in existing audit practices and advocates for designing audits that deliver tangible accountability outcomes. By emphasizing the importance of both internal and external audits, the authors demonstrate how these mechanisms can drive policy reforms, product redesigns, and more comprehensive regulation of AI technologies .

Finally, Novelli et al. [333] offer a structured framework for understanding AI accountability, consisting of seven interrelated features: context, range, agent, forum, standards, process, and implications. Their framework elucidates the relational nature of accountability and its dependence on sociotechnical contexts. By analyzing the goals of compliance, reporting, oversight, and enforcement, Novelli and colleagues provide actionable insights for policymakers aiming to balance ethical, legal, and operational considerations in AI governance.

For those interested in further exploring transparency, Polat Goktas [334] offers a comprehensive bibliometric study on ethics, transparency, and explainability in generative AI systems. This work focuses on their implications in high-stakes decision-making domains such as healthcare and finance, highlighting critical themes such as the necessity of ethical frameworks, transparency, and explainability, particularly amidst the increasing adoption of large language models like ChatGPT. Goktas underscores the pressing need to align generative AI systems with societal values to uphold public trust and promote responsible innovation.

Similarly, Stefan Buijsman [335] critiques existing approaches to transparency, advancing a value-based model as a more comprehensive alternative. Buijsman identifies the shortcomings of process-based and outcome-based transparency models, arguing that these approaches inadequately address the complexities inherent in socio-technical systems. The proposed value-based transparency framework emphasizes the integration of societal and ethical considerations throughout the design and deployment of AI systems, ensuring that transparency not only fosters accountability but also aligns with public legitimacy. This holistic perspective is particularly relevant in the context of public-sector applications, where societal trust and ethical governance are paramount.

Chapter 6

Regulatory Initiatives in AI

6.1 Introduction

Several multilateral initiatives have addressed the ethical implications of AI. In Europe, the most widely publicized initiatives have originated from political decision-making bodies, such as the European Parliament [98]. In contrast, other initiatives have originated from private corporate sectors, including those defined by companies like Amazon [6], Microsoft [203], META [199], Google [121], and OpenAI [232]. Although there are similarities in the definitions and principles these initiatives aim to uphold, there are also notable nuances. European Parliament initiatives have generally aligned with the concept of Trustworthy AI, whereas private corporate initiatives often refer to the idea of Responsible AI. While these concepts are not mutually exclusive, they emphasize different aspects and nuances that reveal the specific priorities of each corporation. We will review the key aspects of several of these initiatives to elucidate the differences and similarities between these approaches.

6.1.1 General Data Protection Regulation (GDPR, 2016)

The GDPR (General Data Protection Regulation) [236] is a regulatory initiative by the European Parliament and the Council of the European Union aimed at protecting natural persons concerning the processing of personal data and the free movement of such data. Enacted in 2016, the GDPR establishes that protecting personal data is a fundamental right. This aspect is crucial for developing intelligent systems, as AI models often rely on personal data. The

GDPR provides a regulatory framework for the use of personal data across various fields, including those involving the creation of AI models [243].

The GDPR recognizes the protection of personal data as a right but states that it is not an absolute right. According to it, data protection should be designed to serve humanity. Hence, data protection must be considered in relation to its societal function and balanced against other fundamental rights. The GDPR upholds fundamental rights, including respect for private life, freedom of expression, freedom of conscience and religion, the right to fair treatment, and respect for cultural, religious, and linguistic diversity.

The GDPR is a comprehensive regulatory framework that governs the use of personal data by both public and private sectors, extending beyond intelligent systems to areas like national security, economics, education, health, and culture. It distinguishes between personal data use in private activities, such as social networking, and professional or commercial contexts. While the GDPR applies to data controllers, it exempts authorities engaged in criminal investigation and public security activities.

The use of personal data by corporations or individuals for purposes other than criminal investigation and public security must involve informed consent. The GDPR broadly defines informed consent as a clear affirmative act through which an individual authorizes the processing of their personal data in a freely given, specific, informed, and unambiguous manner. This includes electronic means as well as oral statements. Affirmative means include checking a box on a website, agreeing to terms of use on data platforms, or other explicit selective affirmative methods. Implicit acceptances, pre-ticked boxes, or other passive acceptance mechanisms are excluded. In the context of scientific research, since the utility of data may not be clear at the outset of research, informed consents authorize use in broad research areas, without the need to specify the exact uses derived from the data provided.

The GDPR mandates that all personal data processing must be fair, meaning its use must be transparent to the individuals from whom the data is collected, used, consulted, or otherwise processed. The transparency principle requires that all information related to the processing of personal data be easily accessible and described in clear, plain language. This principle mainly concerns the purpose for which the data controller processes the data as well as the necessary information to ensure fair use and transparent processing, respecting individuals' rights to access the information. The GDPR also underscores the importance of using personal data for a clearly defined and limited period, requiring necessary measures to omit or correct any inaccurate personal data.

The GDPR principle of transparency stipulates that information accessed or derived from the use of personal data should be public, easily accessible, and easy to understand, and when appropriate, include visualizations. Such information should be accessible digitally, for example, through websites.

This regulatory framework advocates for the safe and transparent handling of processed data concerning individuals, considering various circumstances and the context in which personal data is processed. This involves carefully considering factors that could result in inaccuracies. Data must be used securely in a way that considers potential risks associated with the interests and rights of data subjects, and prevents discriminatory effects on individuals based on race, ethnicity, political stance, religion, beliefs, group membership, health status, sexual orientation, or any outcomes that might arise from profiling.

The GDPR imposes obligations on controllers in areas involving the use of personal data for making automated decisions at the individual level. Individuals have the right to object to the use of their data, particularly when decisions are made based on profiles created from their data. Controllers are required to demonstrate legitimate and compelling reasons for the use of personal data. Individuals have the right not to be subject to a decision based solely on automated processing, including profiling, which produces legal or similarly significant effects.

The GDPR defines the role of the data protection officer, who is appointed by the controller to respond to data transparency requests, unless the request is judicial. This officer is responsible for overseeing data processing monitoring operations and ensuring compliance with the GDPR's regulatory framework.

Additionally, the GDPR outlines the need for certification mechanisms at the state level. Alongside this, it mandates the establishment of accrediting agencies. These agencies are involved with the accountability and transparency aspects of each initiative related to the use of personal data. The supervisory and investigative action, defined as the forum of the accountability relationship, is termed an audit.

6.1.2 AI audits: The early initiatives

One of the earliest audit practices was initiated by the UK's Information Commissioner's Office (ICO) to ensure organizations comply with data protection laws, particularly the GDPR. The ICO conducts both consensual and compulsory audits to assess personal data processing and provides guidance on improvements. These audits aim to raise awareness of data protection, demonstrate organizational commitment, and build public trust, fostering innovation and growth.

The UK adopted the GDPR through the Data Protection Act 2018 (DPA 2018), which governs how personal information is used by organizations, businesses, and the government, ensuring that data is handled legally and ethically. As the UK's counterpart to the EU's GDPR, the DPA 2018 plays a crucial role in regulating personal data across sectors, protecting individuals from unauthorized or improper use of their data.

The DPA 2018 sets out strict data protection principles that all entities handling personal data must follow. These principles ensure data is processed fairly, lawfully, transparently, and for specific, stated purposes. They also require that data processing be adequate, relevant, and limited to what is necessary, while ensuring data accuracy, timely updates, and secure retention only for as long as needed. To safeguard data, appropriate security measures must prevent unlawful or unauthorized processing, access, loss, or damage.

Individuals are granted significant rights under the DPA 2018, including the right to access and correct their data, request data erasure under certain conditions, and restrict or object to data processing. The act also provides enhanced protections for sensitive data, such as racial or ethnic origin, political opinions, and health information. Additional safeguards are in place for data related to criminal convictions and offences, as well as rights concerning automated decision-making and profiling.

The ICO's audit process starts with an introductory meeting to define the audit's scope and methodology, tailored to each organization's specific risks and concerns. The audit scope typically includes various operational and management areas such as governance, contracts, training, and data minimization. The audit involves reviewing relevant documents, interviewing key personnel, and directly observing operational practices.

Practical aspects of the audit include minimal disruption to daily operations, with the ICO making use of remote auditing techniques where appropriate. On-site activities might still be necessary for thorough inspections. Throughout the audit, the ICO's team interacts with the organization's staff to understand and evaluate the implementation and effectiveness of data protection measures.

The ICO will identify and prioritize high-risk data controllers for audits based on multiple criteria. These include the volume and nature of reported breaches, complaints received by the ICO, and the content of the controllers' annual statements concerning their data control practices. Additional sources of information such as media reports and other publicly available data will also inform the selection process. The potential impact of non-compliance is evaluated by considering how many individuals are affected, the sensitivity of the processed data, and the extent of harm or distress that non-compliance might cause. This comprehensive approach allows the ICO to target its resources effectively towards organizations where the risk of data mishandling is greatest.

Organizations selected for audits are categorized into five groups: voluntary organizations, those identified through risk assessments, entities providing relevant educational opportunities in areas of particular interest to the ICO, organizations recommended by other ICO departments for an audit, and those identified through investigations by the ICO's Investigations Team. This classification ensures that the audit program is both targeted and adaptive to evolving risks and areas of concern within data protection.

Throughout the audit, regular communication with key staff helps assess operational effectiveness, supplemented by data analysis and control testing. If a data breach occurs, immediate steps are communicated. Daily updates on areas of concern are provided, culminating in a closing meeting where major issues and further steps are discussed. Finally, a draft report is issued for the organization to review and create an action plan, followed by the delivery of a final report and executive summary.

6.1.3 Artificial Intelligence High-Level Expert Group (AI HLEG - 2018)

The GDPR serves as a broad regulatory framework for protecting and managing personal data, with significant implications for AI. Given AI's importance, the European Parliament tasked a group of experts with developing an AI Strategy for Europe. This initiative, which included members from industry, academia, and civil society, aimed to ensure diversity, coherence, and consistency in Europe's approach to AI.

The group, known as the AI High-Level Expert Group (AI HLEG) [294], released guidelines in December 2018 for a **Trustworthy AI** [140]. According to the document, a Trustworthy AI should adhere to existing laws and regulations, uphold ethical principles and values, and demonstrate robustness both technically and in its societal impacts.

In April 2019, the group published ethical guidelines for a Trustworthy AI. The document outlines a framework by defining its fundamentals and then detailing an action plan towards achieving it. The foundations emphasize the importance of developing, deploying, and utilizing AI systems that respect ethical principles such as human autonomy, harm prevention, fairness, and explainability. It also highlights the need to recognize and address potential tensions among these principles and to pay special attention to scenarios affecting vulnerable groups, such as children, people with disabilities, and those historically disadvantaged or at risk of exclusion. Furthermore, the guidelines stress the need to acknowledge power or information asymmetries, such as between employers and workers or between businesses and consumers. The document also notes that while AI systems can offer substantial benefits to individuals and society, they also pose certain risks and can have adverse effects, some of which might be hard to predict, identify, or measure. For example, impacts on democracy, the rule of law, distributive justice, or the human mind itself. Therefore, appropriate measures should be taken to mitigate these risks, with the severity of the measures proportionate to the level of risk.

The action plan in the document outlines seven requirements that AI systems must meet to achieve Trustworthy AI. These guidelines include ensuring that the development, deployment, and use of AI systems adhere to the standards for reliable AI, which are: (1) human agency and oversight, (2) technical robustness and safety, (3) privacy and data governance, (4) transparency, (5) diversity, non-discrimination and fairness, (6) societal and environmental well-being, and (7) accountability. To ensure these requirements are met, it will be necessary to consider employing both technical and non-technical methods. There should also be a push to enhance research and innovation to assess AI systems and promote compliance; results and open interpretation questions should be publicly disclosed, and a new generation of AI Ethics specialists should be systematically trained. The need to clearly and proactively communicate information to stakeholders about the capabilities and limitations of AI systems is emphasized, facilitating the setting of realistic expectations and understanding of compliance with the requirements. Transparency about using an AI system, ensuring traceability and auditability, especially in critical contexts or situations, is crucial. There is also a need to involve stakeholders throughout the AI system lifecycle, promoting education and training so that all parties are aware of Trustworthy AI and receive appropriate instruction. Finally, it is vital to acknowledge that fundamental tensions may exist between different principles and requirements; these tensions and their resolutions must be consistently identified, evaluated, documented, and communicated.

The third document published by the group in July 2020 addressed the necessary requirements for achieving Trustworthy AI. These are detailed in the "Glossary of Assessment List for Trustworthy Artificial Intelligence (**ALTAI**)" [139] and essentially expands upon the requirements outlined in the ethical guidelines document for Trustworthy AI. These include:

- **Human Agency and Oversight:** AI systems should support human action and decision-making, adhering to the principle of respecting human autonomy. This requires that AI systems act as enablers for a democratic, flourishing, and equitable society and uphold fundamental rights underpinned by human oversight.

- **Technical Robustness and Safety:** A critical requirement for achieving reliable AI systems is their trustworthiness (the ability to provide services that can be justifiably trusted) and resilience (robustness when facing changes). Technical robustness demands that AI systems be developed with a preventative approach to risks and that they function reliably as intended, while also minimizing unintended and unexpected harms and preventing them when possible. This applies especially in scenarios involving potential changes in their operational environment or interactions with other agents (human or artificial) that may adversely affect the AI system. The requirement encompasses four aspects:

1. ***Security:*** This concerns the possibility of the intelligent system causing harm to humans.
2. ***Safety:*** This relates to the use of metrics and assessments of risk levels during the development of the intelligent system.
3. ***Accuracy:*** This involves the relationship between harm and low accuracy levels in the system.
4. ***Reliability, fallback plans, and reproducibility:*** They pertain to risks and damages caused by the system due to unreliability or non-reproducible outcomes.

- **Privacy and Data Governance:** Closely linked to the harm prevention is privacy, a fundamental right particularly impacted by AI systems. Preventing harm to privacy necessitates proper data governance, covering the quality and integrity of the data used, its relevance given the domain in which the AI systems will be implemented, its access protocols, and the ability to process it in a way that protects privacy. This requirement is linked to the GDPR.

- **Transparency:** A crucial component for achieving Trustworthy AI is transparency, which includes three elements:

  1. ***Traceability:*** This involves assessing whether the development processes of the AI system, i.e., the data and processes generating the system's decisions, are adequately documented to enable traceability, enhance transparency, and ultimately build trust in AI within society.
  2. ***Explainability:*** This is the ability to explain both the technical processes of the AI system and the reasoning behind the decisions or predictions it makes. Explainability is vital for generating and maintaining user trust in AI systems. AI-driven decisions should be explained and understood, as much as possible, by those directly and indirectly affected to enable the challenging of such decisions.
  3. ***Communication about limitations:*** This involves assessing whether the capabilities and limitations of the AI system have been communicated to users appropriately for the use case at hand. This could include communicating the accuracy level of the AI system, as well as its limitations.

- **Diversity, Non-discrimination, and Fairness:** To achieve Trustworthy AI, we must foster inclusion and diversity throughout the AI system's lifecycle. AI systems—both during training and operation—may incorporate unintended historical biases, incomplete models, and poor governance. The perpetuation biases could lead to direct prejudice and discrimination against specific groups or individuals, potentially exacerbating bias and marginalization. Harm can also arise from the intentional exploitation of biases or engaging in unfair competition, such as price homogenization through collusion or a non-transparent market. Wherever possible, identifiable and discriminatory biases should be removed during the data collection phase. AI systems need to be user-centric and designed to allow all individuals to use AI products or services, regardless of age, gender, abilities, or characteristics. Ensuring accessibility for people with disabilities, who are present in all social groups, is particularly crucial. This aspect is closely tied to data bias and algorithmic bias, both of which involve how data-driven intelligent systems can collect data that perpetuates biases, thereby reproducing stereotypes and widening gaps in unequal treatment affecting individuals or groups in society. Additionally, AI systems must not adopt a one-size-fits-all approach and should embrace the principles of Universal Design [82], addressing the broadest range of users and adhering to relevant accessibility standards. This will enable equitable access and active participation of all people in both existing and emerging computing systems.

- **Societal and Environmental Well-being:** In line with the principles of fairness and harm prevention, society at large, other sentient beings, and the environment should be considered stakeholders throughout the lifecycle of AI systems. The pervasive exposure to social AI systems in all areas of our lives may negatively impact our social relationships and attachments. While AI systems can enhance social skills, they may also contribute to their deterioration. This could affect the physical and mental well-being of individuals. Therefore, monitoring and carefully considering the effects of AI systems is essential. Additionally, sustainability and ecological responsibility should be promoted, and research into AI solutions addressing areas of global concern should be encouraged. In general, AI should be used to benefit all human beings, including future generations. AI systems should serve to maintain and promote democratic processes and respect the plurality of individuals' values and life choices. AI systems must not undermine democratic processes, human deliberation, or democratic voting systems, nor pose a systemic threat to society at large. Special emphasis is placed on this last point. It will be necessary to self-assess the impact of an AI system from a social perspective, considering its effect on institutions, democracy, and society at large, for example, when AI systems exacerbate fake news, segregate the electorate, or facilitate totalitarian behavior.

- **Accountability:** The principle of accountability mandates the establishment of mechanisms to ensure responsibility for the development, deployment, and/or use of AI systems. This issue is closely linked to risk management, involving the identification and mitigation of risks in a transparent manner that can be explained and audited by third parties. Should unjust or adverse impacts occur, there needs to be accessible accountability mechanisms in place to ensure a proper opportunity for redress. This requirement emphasizes AI auditability, which involves the self-assessment of the current or required level for evaluating the AI system by both internal and external auditors. The ability to conduct evaluations and access to these evaluations can contribute to a Trustworthy AI. In applications that affect fundamental rights, including those critical to safety, AI systems should be designed to allow independent audits.

The document outlines several fundamental concepts. For instance, it defines accountability as "the idea that one is responsible for their actions—and, consequently, their outcomes, and must be able to explain their goals, motivations, and reasons." Accountability possesses multiple dimensions and is sometimes mandated by law. For example, the GDPR requires organizations handling personal data to implement security measures to prevent data breaches and mandates reporting if these measures fail. However, accountability can also reflect an ethical standard that may not necessarily lead to legal consequences. Certain tech companies may choose not to develop facial recognition technology, despite there being no ban or technological moratorium, based on ethical considerations of accountability.

The document also defines fairness as encompassing a range of concepts known as equity, impartiality, egalitarianism, non-discrimination, and justice. Fairness is primarily concerned with the ideal of equal treatment among individuals or groups, often referred to as 'substantive' fairness. Additionally, fairness includes a procedural aspect, which involves the ability to seek and obtain redress when individual rights and freedoms are infringed.

6.1.4 EU regulatory framework on AI (EU AI Act)

Subsequent efforts by the European Parliament aimed to design a regulatory strategy for establishing limits and responsibilities in the deployment of Trustworthy AI. On April 21, 2021, the document "Laying down harmonized rules on AI and amending certain union legislative acts" [61] was released. This document includes an explanatory memorandum that accompanies the proposed regulation. The memorandum outlines contextual factors that justify the promotion of the proposal. It is stated that the need arises because AI can support socially and environmentally beneficial outcomes and provide competitive advantages to companies, thereby benefiting the European economy. These acts impact significant sectors such as climate change, health, environment, public sector, and agriculture. Despite its benefits, it is acknowledged that AI poses risks, which necessitate a balanced approach to AI regulation.

The need for defining regulation stems from the European Commission's white paper "A union that strives for more", which highlights the need for legislation for a coordinated approach on the ethical implications of AI. Following this announcement, on February 19, 2020, the European Commission published the white paper "A European Approach to Excellence and Trust." This paper discusses promoting the development of AI while simultaneously considering its risks. Additional context, such as the conclusions from the European Council on October 21, 2020, highlights the necessity to tackle the opacity, complexity, bias, and level of autonomy of certain AI systems to ensure they align with fundamental rights and comply with the legal framework. The European Council also provided background for the proposal based on resolutions related to AI, addressing issues concerning ethics, copyright, and other privacy-related matters within the GDPR framework. Drawing from all these contexts, the European Commission proposes an AI regulatory framework with the following objectives:

- Ensuring that AI systems are safe and comply with current legal frameworks,
- Securing legal certainty to facilitate investment and innovation in AI,
- Improving governance and effective enforcement of existing laws based on the fundamental rights applicable to AI systems,
- Facilitating the development of trustworthy AI applications to prevent market fragmentation.

The memorandum sets forth a key principle for the design of the regulatory proposal:

A key principle for the design of the regulatory proposal

Regulatory efforts should be confined to the minimum requirements necessary to address risks associated with AI without stifling technological advancement.

The challenge is creating a forward-looking proposal that remains relevant by anticipating future technological challenges. Additionally, it should clearly define a balanced and proportional regulatory framework focused on well-defined risks to avoid unnecessary restrictions. In essence, it must be a precise framework with clear boundaries and flexible mechanisms that do not hinder AI development.

The proposal establishes rules for the development, market positioning, and use of AI systems based on a risk-based approach. It bans certain harmful practices and sets forth some exceptions, including specific uses of remote biometric identification systems for law enforcement. A key aspect of the proposal is defining what constitutes a **high-risk AI system**. Such systems must adhere to well-established standards within the proposal to be operational. Various obligations also fall on both providers and users of these systems to ensure their safe use. For some specific AI systems, only minimal transparency obligations are defined, particularly those involving chatbots and deep fakes.

Another fundamental element of the proposal is the definition of a governance scheme. The governance for this regulatory framework stipulates participation by EU member states, outlining mechanisms for cooperation through the **European AI Board**.

The legal basis of the proposal rests on the need to promote the functioning of the internal market, which, due to the magnitude of the challenge, cannot be effectively addressed by individual states independently. The proposal's objectives are set at the EU level to prevent the proliferation of conflicting local regulations, which would create legal uncertainty and hinder investment in AI development.

Reflecting all the efforts led by the EU, as shown in the HLEG and ALTAI initiatives, the AI Alliance was established, consisting of 4,000 stakeholders. This group convenes to discuss the societal and technological implications of AI, culminating in an annual assembly. The proposal is considered to be participatory, engaging, and inclusive, involving stakeholders from various sectors connected to AI. Based on the deliberations, it was decided to develop a Horizontal EU legislative instrument that follows a proportionate risk-based approach, along with codes of conduct for non-high-risk AI systems. The central aspect of the proposal is to outline the risk-based approach and establish general guidelines for the development and use of non-high-risk systems, thereby creating a comprehensive framework for AI operations.

6.1.5 Foundational elements of the EU AI Act

The proposal outlines a list of AI practices that are to be prohibited. These practices are described in Title II, Article 5:

1. Marketing, whether through service offerings or the use of an AI system, that is based on subliminal techniques aimed at distorting a person's behavior to their detriment, either through material or psychological harm.
2. Marketing, whether through service offerings or the use of an AI system, that exploits the vulnerabilities of a specific group based on age or physical or mental disabilities in a way that distorts their behavior and causes material or psychological harm.

3. Marketing, whether through service offerings or the use of an AI system, that relies on public sphere decisions evaluating or classifying an individual's reliability with a social score, which can result in detrimental or unfavorable treatment of individuals or groups in contexts unrelated to the original data collection or lead to disproportionate or unfair treatment based on their social behavior.
4. The use of real-time remote biometric identification systems in public spaces, unless strictly necessary. Necessary use is defined as targeting a person who is potentially a victim of a crime, including missing children; preventing threats to personal integrity, such as from terrorist acts; and the detection, location, identification, and prosecution of perpetrators or suspects of punishable criminal acts.

The list of prohibitions is short but significant. The first two prohibitions address the asymmetry that occurs between the system and its users, whether due to the use of subliminal strategies or the imbalance between the AI system's content generation capabilities and the users' capacity to process these contents, for example, due to a disability or vulnerability. We will further explore these issues in the upcoming chapters, as they are fundamental to what we know as disinformation. Our focus will be on uncovering the relationship between disinformation and AI and how AI systems can exacerbate this phenomenon.

The third prohibition addresses the use of data across different contexts. In AI, this is somewhat related to transfer learning, which involves applying an AI model in domains that differ from those in which it was originally trained. While the prohibition does not target the use of data in different contexts, it specifically addresses the creation of social scores that support decision-making based on social trustworthiness. It is understood that an application case could involve using data from, for example, social networks to infer a person's illness. This information should not be used in a different context, such as analyzing access to medical insurance. Cross-domain inference to support decision-making, especially in the public sphere, is a controversial use of data that even impacts personal privacy. The fourth prohibition essentially establishes a framework for using real-time biometric systems, based on the concept of necessary use.

Another key aspect of the proposal involves classifying high-risk AI systems. These definitions, included in Title III of the proposal, characterize a high-risk AI systems based on the areas in which they are used. These areas include:

- Biometric identification and classification of natural persons.
- Management and operation of critical infrastructure.
- Education and vocational training.
- Employment, human capital management, and access to self-employment.

- Personnel selection, credit evaluation, and prioritization systems for emergency response.
- Systems that provide support for the judicial system.
- Border control and migration systems.
- Support systems for democratic processes.

The proposal addresses the definition of duties and requirements that high-risk AI systems must meet. It specifies that the system should be implemented, documented, and maintained recognizing its high-risk status. In this context, it specifies that a high-risk AI system must incorporate a continuous and iterative process throughout its entire lifecycle, addressing:

1. The identification and analysis of known and potential risks associated with the system;
2. The estimation and evaluation of risks that may emerge when the high-risk AI system is utilized, both under its intended conditions and other scenarios;
3. The assessment of emerging risks based on post-market monitoring data analysis, and
4. The adoption of risk mitigation measures.

The requirements specify a post-market monitoring system that must systematically gather information from users about the system's performance throughout its lifecycle. Once the risks are defined, it is necessary to ensure the elimination or reduction of risks both in the design and implementation of the system. When necessary, control and mitigation measures should be implemented for risks that cannot be eliminated. Additionally, users must be informed about these risks. Regarding this final point, the information provided to users must support transparent system operation enabling them to interpret the system's results and use it appropriately. This includes instructions for use, the system's purpose, its accuracy level, robustness, and cybersecurity, its performance concerning specific groups it targets, and, when applicable, specifications related to the data used for training, validation, and testing.

Another crucial aspect of the proposal concerns data and data governance. It is established that the data partitions for training, validation, and testing must be subject to appropriate data governance and management practices. These practices should cover various stages of the data lifecycle, including collection, data annotation, labelling, cleaning, enrichment, and aggregation, a rigorous review based on potential biases and identification of possible data gaps or shortcomings in the data, and how these gaps can be addressed.

The proposal also defines the importance of human oversight. This oversight should prevent or minimize risks to health, safety, or fundamental rights arising when a high-risk AI system is used. Human supervision must be ensured before the system is released to the market.

The proposal outlines all obligations assumed by the providers and all parties involved in the lifecycle of the high-risk AI system. These parties include competent authorities, product manufacturers, importers, distributors, and users. Each actor has a set of obligations established in the proposal.

To ensure compliance with the established requirements and duties, the proposal refers to standards and conformity assessments. It emphasizes the need to define harmonized standards, which must align with the requirements specified in the proposal. If a provider adopts these harmonized standards, they must follow a conformity assessment procedure based either on internal control strategies or on a quality assurance system. The **quality assurance system** should include the following elements:

1. A strategy for regulatory compliance, including conformity procedures.
2. Techniques, procedures, and systematic actions to be used for the design and verification of a high-risk AI system.
3. Techniques, procedures, and systematic actions to be employed in the development, quality control, and quality assurance of high-risk AI systems.
4. Examination, testing, and validation to be conducted before, during, and after the system's development.
5. Technical specifications, including standards to be applied and methods to ensure compliance with the proposal's objectives when harmonized standards are not fully applied.
6. Systems and procedures for data management, covering all data lifecycle stages.
7. The risk management system.
8. The implementation and maintenance of the system after it enters the market.
9. Procedures related to the reporting of serious incidents.
10. Communication management with competent authorities.
11. Systems and procedures for documenting all relevant information.
12. Resource management, including measures related to the security of supply.
13. An accountability framework outlining the responsibilities of management and other staff.

The proposal also sets out transparency obligations for other AI systems (Title IV). Notably, it requires providers to ensure that AI systems interacting with individuals are designed so that these individuals are informed when they are interacting with an AI system unless it is obvious from the context of use. This obligation does not apply to systems legally authorized for detecting and preventing criminal activities. Another transparency-related obligation mandates that users of emotion recognition systems or biometric categorization systems inform individuals that they are subject to these systems. This does not apply to specific biometric categorization systems authorized by law for detecting and preventing criminal activities. Furthermore, users of an AI system that generates or manipulates images, audio, or videos resembling people, objects, places, or other entities in a way that could misleadingly appear as authentic content ('deep fake') must disclose that the content has been artificially generated or manipulated.

Finally, the proposal establishes the governance of the regulation, headed by the European Artificial Intelligence Board, which is tasked with assisting and advising the European Commission on cooperation and coordination aspects necessary for implementing the regulatory framework. It also defines the National Competent Authorities, which must maintain the confidentiality of the information and data obtained during the enforcement of the regulation. The imposition of penalties and administrative sanctions falls to the EU member states. Penalties must be effective, proportionate, and dissuasive. They should consider the interests of small-scale providers and startups in a way that does not compromise their economic viability. Economic sanctions can amount to up to 30 million euros or, if the offender is a company, up to 6% of its total worldwide annual turnover for the preceding financial year, whichever is higher, if there is non-compliance with the prohibition of AI practices or data and data governance safeguards. Non-compliance with other obligations defined in the regulation will be subject to economic sanctions of up to 20 million euros or, if the offender is a company, up to 4% of its total worldwide annual turnover for the preceding financial year. Misleading information also carries penalties of up to 10 million euros or, if the offender is a company, up to 2% of its total worldwide annual turnover for the preceding financial year.

At the time this book was written, the final version of the AI Act had been published in December 2023 [98] and encompassed most aspects initially outlined in the extensive discussions on the foundational principles of the legislation, which are summarized in this section. This broad discussion, which led to the white paper on the AI Act [61], went through a consolidation phase, resulting in the EU AI Act. The law represents one of the first transnational regulatory frameworks in AI.

6.2 AI audits and governance

The increasing risks associated with AI have prompted institutions to develop frameworks that adhere to the principles of a responsible AI. The necessity to demonstrate and certify adherence to the principles of responsible AI has led to the development of AI audit initiatives. These audits offer various methods for examining AI sociotechnical systems that may have a social impact. AI audits must document these impacts and how they occur.

AI audits primarily focus on verifying developers' adherence to principles of transparency, fairness, and accountability throughout the AI lifecycle, as well as compliance with legislation, regulation, and other guidelines derived from public policies. As illustrated by the EU regulatory proposal, which defines prohibitions and a list of requirements for the development and use of high-risk systems, it is increasingly becoming essential to verify compliance with this kind of requirements. AI audit initiatives aim to verify the compliance of processes and outcomes within an existing regulatory framework.

AI audits are supported under the umbrella of AI governance. Various AI governance institutions provide documents with recommendations or standards to which developers must adhere. This is the case with the EU Parliament and the EU AI Act. Alongside this initiative, other institutions have joined these efforts.

6.2.1 US-based Government Accountability Office (GAO)

The GAO developed an accountability framework for federal agencies and other entities to ensure accountability and responsible use of AI in government programs and processes. The framework is organized around four principles that address governance, data, performance, and monitoring of AI systems. For each principle, the framework outlines key practices for federal agencies, other institutions, and AI system developers. It includes essential practices, questions for developers, and procedures for auditors.

Governance: The framework defines governance as the management and oversight of AI by those in charge, who can utilize a governance structure to manage risk, demonstrate integrity and adherence to ethical values, and ensure compliance with relevant laws and regulations. The framework identifies nine practices which are grouped at both the organizational and system levels. At the organizational level, managers are expected to establish an environment that fosters a positive and proactive attitude towards internal control. There are six key practices that help establish these principles at the **organizational level**:

1. **Clear objectives:** Define clear goals for the AI system to ensure that the expected outcomes are achieved.
2. **Roles and responsibilities:** Define clear roles along with responsibilities to ensure effective operations, timely corrections, and oversight.
3. **Values:** Demonstrate commitment to the values and principles established by the entity to promote public trust in the responsible use of AI.
4. **Workforce:** Recruit, develop, and retain personnel with multidisciplinary skills and experiences in the design, development, and monitoring of AI systems.

5. **Stakeholder involvement:** Include diverse perspectives from the stakeholder community throughout the AI lifecycle to mitigate risks.
6. **Risk management:** Implement a risk management plan to systematically identify, analyze, and mitigate risks.

At the system level, governance advocates for AI systems' compliance with the regulatory framework. Unlike the organizational level, which focuses on the organization's practices and processes, system governance targets the AI system's compliance within the context in which it will be used, as defined by the relevant regulations and laws. In this regard, the framework establishes three recommended practices:

1. **Specifications:** Establish and document technical specifications to ensure that the AI system fulfills its purpose.
2. **Compliance:** Ensure that the AI system complies with relevant laws, regulations, standards, and guidelines.
3. **Transparency:** Promote transparency to allow external stakeholders to access information on the AI system's design, operation, and limitations.

The framework provides a list of questions to consider and a set of audit procedures for each practice. These elements are crucial for operationalizing the framework. Figures 6.1 and 6.2 show examples of questions and audit procedures for each governance practice.

Data: The framework specifies that the data used to train, validate, and test an AI system must be suitable for the purpose of ensuring that the system produces consistent and accurate results. Data management must adequately ensure the quality, reliability, and representativeness of the data. To fulfill this purpose, the framework distinguishes between data used for model development and data used for system operation. The suggested practices for system development include:

1. **Sources:** Document the origin of the data sources used to train the system (data provenance).
2. **Reliability:** Data used to train a system must be reliable, as data reliability affects the accuracy of the results.
3. **Categorization:** Define the attributes used to categorize data, documenting the rationale used for organizing the data and how it was segmented into training, validation, and testing partitions.
4. **Variable Selection:** Define the variables used to construct each component of the AI system.
5. **Enhancement:** Define the use of synthetic, imputed, or augmented data.

Concerning the data used for system operations, the framework outlines the following practices:

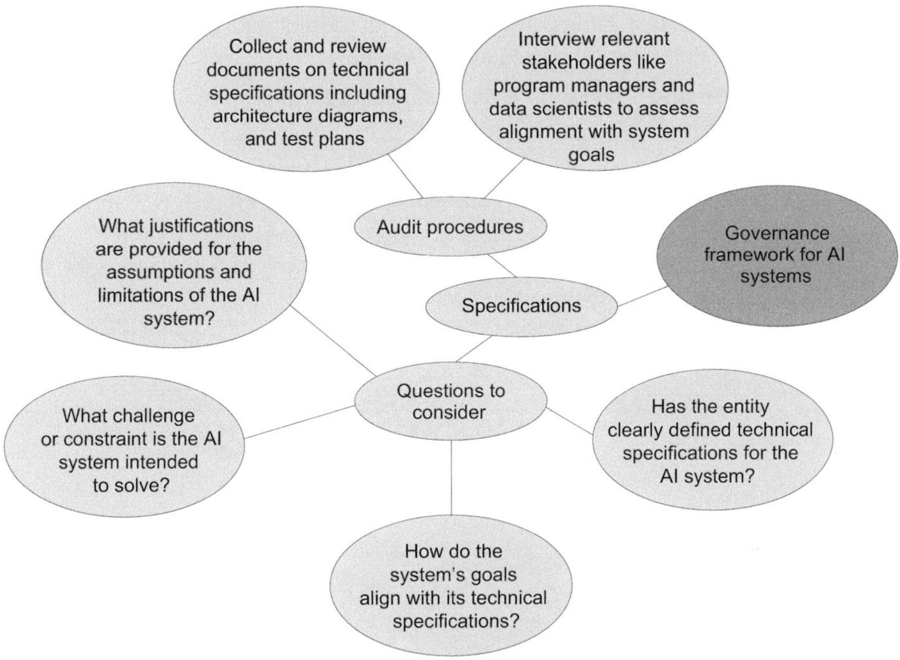

Figure 6.1: Audit procedures and questions to consider regarding specifications in AI audits.

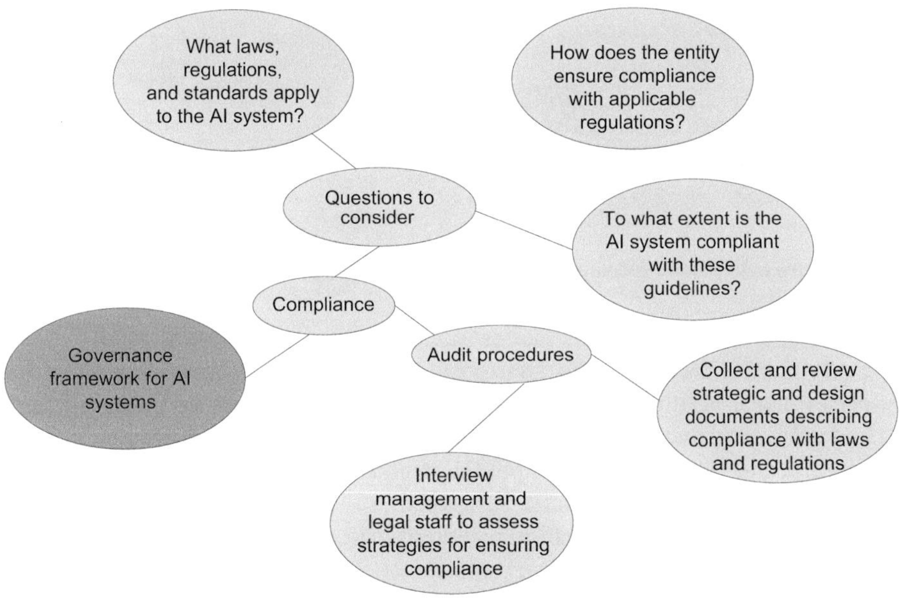

Figure 6.2: Audit procedures and questions to consider regarding compliance in AI audits.

1. **Dependencies:** Define the interconnectivities and dependencies of data streams that operationalize the AI system.
2. **Security and Privacy:** Define data security and privacy measures for the AI system.

Similar to governance, the framework outlines questions and audit procedures for each defined data practice. A summary is shown in Figures 6.3 and 6.4.

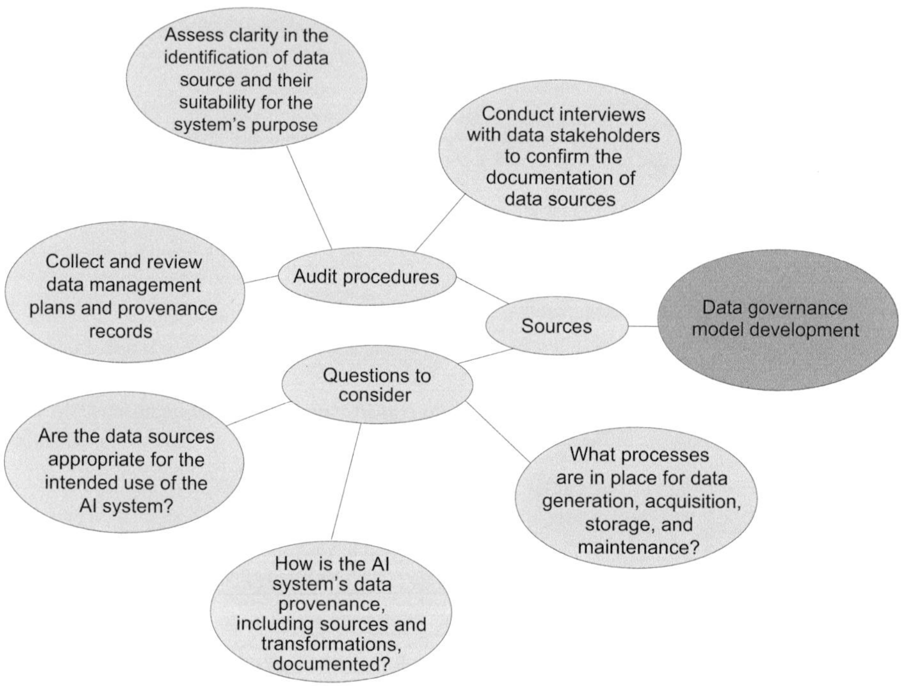

Figure 6.3: Audit procedures and questions to consider regarding sources in AI audits.

Performance. Management and those in charge of AI supervision should use performance verification methods to improve the accuracy of model outcomes and processes, thereby facilitating decision-making for supervisors or corrective actions and contributing to public accountability. In this dimension, the framework defines nine practices grouped at the component and system levels. At the component level, performance verification of each component is defined, as components are the building blocks of AI systems. At the system level, performance verification determines which components operate properly as an integrated whole. The component-level practices include:

1. **Documentation:** Catalog of models and components that do not include models, along with operational specifications and parameters.

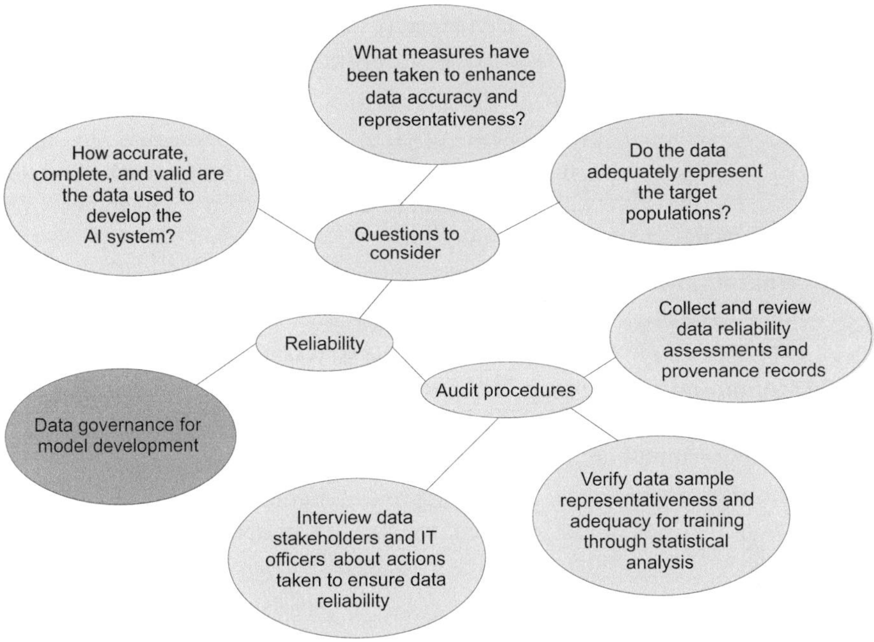

Figure 6.4: Audit procedures and questions to consider regarding reliability in AI audits.

2. **Metrics:** Performance metrics used during the system development phase, which must be accurate, consistent, and reproducible. Metrics should extend beyond accuracy and include bias, equity, and other societal considerations. They should reflect the societal impact expected of the AI system.
3. **Verification:** Verify the performance of each component in relation to the defined metrics to ensure the system functions as expected.
4. **Outputs:** Verify which outputs from each component are appropriate for the operational context of an AI system.

A summary of the main questions and procedures related to AI audit performance at the component level is shown in Figures 6.5, 6.6, 6.7 and 6.8. System-level practices include:

1. **Documentation:** Document verification methods, performance metrics, and outputs of the AI system to provide transparency.
2. **Metrics:** Define metrics to be used for system-level evaluation.
3. **Verification:** Verify performance in relation to the defined metrics to ensure that the AI system is robust against any attempt to misuse the system.
4. **Bias:** Identify potential biases, inequities, and other societal concerns resulting from the AI system.

5. **Human Supervision:** Define and develop procedures for human supervision of the AI system that ensure accountability.

The framework emphasizes the importance of human supervision in AI systems. A detailed presentation of the various approaches to human supervision is found in the "Model AI Governance Framework, 2nd Ed." developed by the Personal Data Protection Commission (Singapore) [328]. These approaches are categorized into three:

1. **Human-in-the-loop:** This approach involves active human supervision of the AI system, where the human retains full control over decision-making, and the AI system only provides inputs or recommendations.
2. **Human-out-of-the-loop:** This refers to the absence of human supervision in decision execution, where the AI system has full control over decision-making without human oversight.
3. **Human-on-the-loop:** In this approach, human supervision intervenes when the outcomes provided by the AI system are not as desired. Otherwise, the system operates without human supervision.

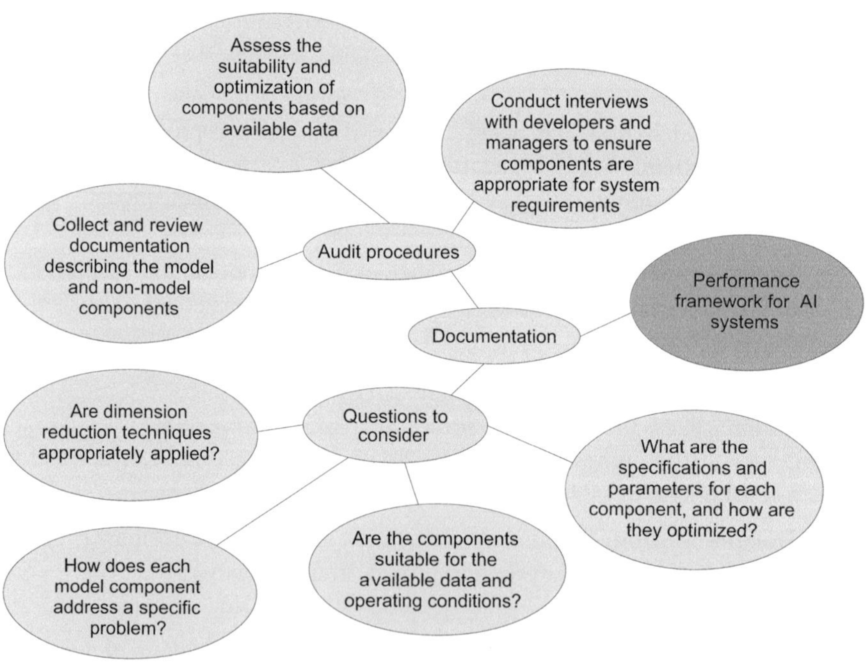

Figure 6.5: Audit procedures and questions to consider regarding documentation in AI audits at the component level.

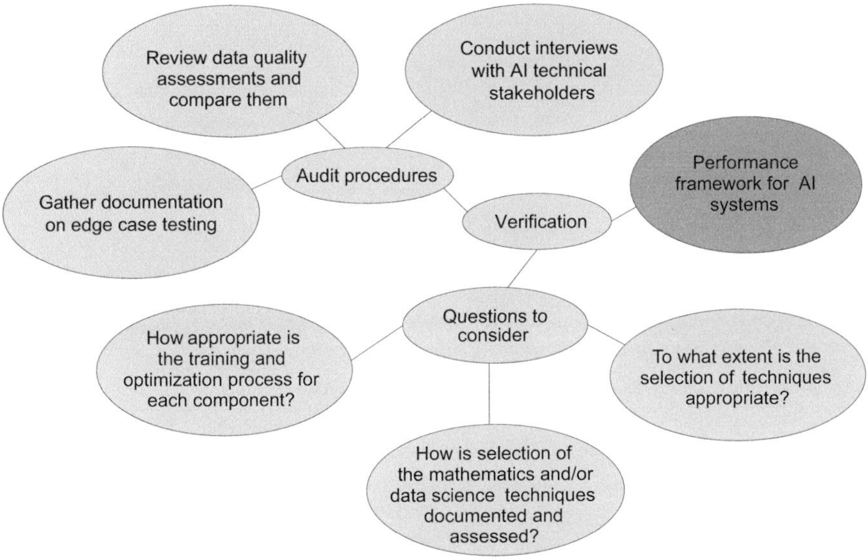

Figure 6.6: Audit procedures and questions to consider regarding verification in AI audits at the component level.

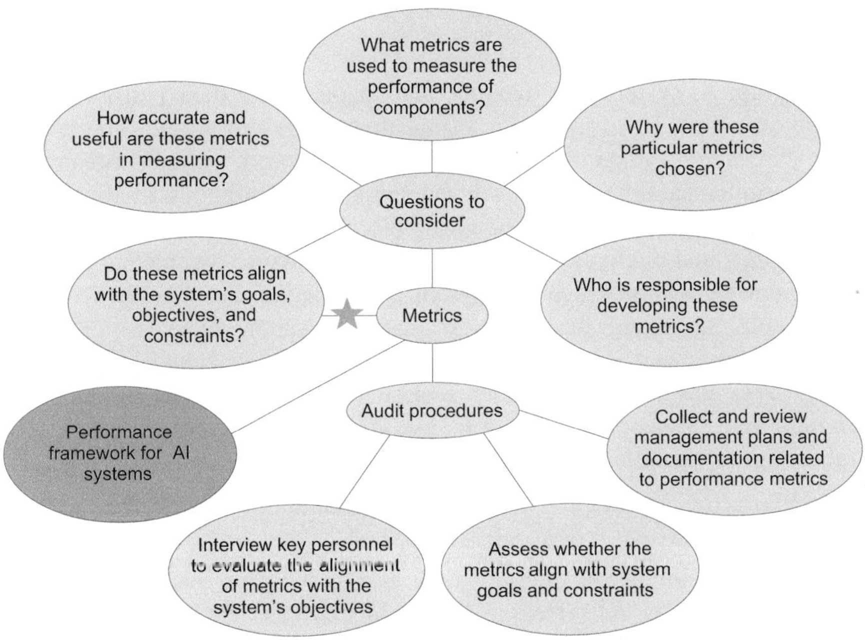

Figure 6.7: Audit procedures and questions to consider regarding metrics in AI audits at the component level.

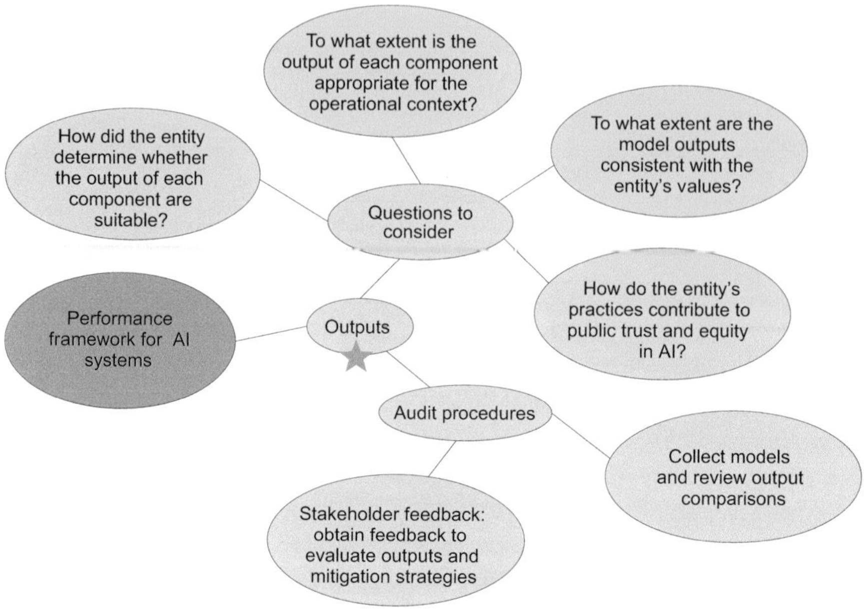

Figure 6.8: Audit procedures and questions to consider regarding outputs in AI audits at the component level.

Monitoring. AI systems are dynamic and adaptive, and their performance can change over time. It is essential to establish a monitoring framework to ensure that the AI system maintains its utility. This framework outlines five practices for this principle, divided into the categories of continuous monitoring and sustainable verification. Continuous monitoring involves tracking input data, outputs generated by predictive models, and performance parameters that define the framework within which the system must operate. Sustainable verification focuses on examining the utility of the AI system, especially when the regulatory framework and operational environment may change over time. In some cases, entities may consider scaling the use of the AI system across geographic locations or expanding its use in different operational settings. The practices grouped under continuous monitoring include:

1. **Planning:** Developing plans for the continuous monitoring of the AI system to ensure it performs as expected.
2. **Drift:** Defining the range of data and model drift that is acceptable to ensure that the AI system produces the expected results. Data drifts refer to changes in the statistical properties of the input data in an operational environment compared to the data used for training. Model drift refers to changes in the relationships between data inputs and prediction outputs. Both data and model drifts can degrade a system's performance, and a tolerable range of drifts must be defined.

3. **Traceability:** Documenting the results of monitoring activities and corrective actions taken to promote system transparency.

A summary of the main questions and procedures for AI auditing related to monitoring is presented in Figures 6.9, 6.10, and 6.11. Regarding the dimension of sustainability and expansion of use, the framework defines two practices:

1. **Ongoing assessment:** Verifying the system's utility to ensure its relevance to the context in which it is used. Drastic context changes, such as the use of AI systems during the COVID-19 pandemic, require performance reviews to match the new context.
2. **Scaling:** Identifying conditions, if any, under which the AI system can be scaled or expanded beyond its usual use.

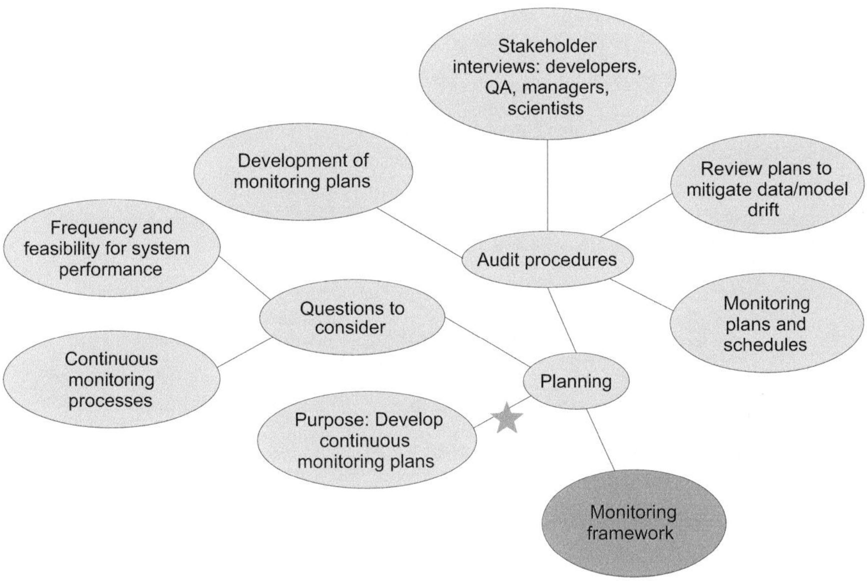

Figure 6.9: Audit procedures and questions to consider regarding monitoring and planning in AI audits.

6.2.2 US Artificial Intelligence Risk Management (US NIST AI RMF)

In January 2023, the National Institute of Standards and Technology (NIST) of the US Department of Commerce released the AI Risk Management Framework (AI RMF) to assist organizations in mitigating the risks of AI systems. These risks surpass those addressed in traditional software regulations. The framework highlights that unlike conventional software, AI systems are trained on data that

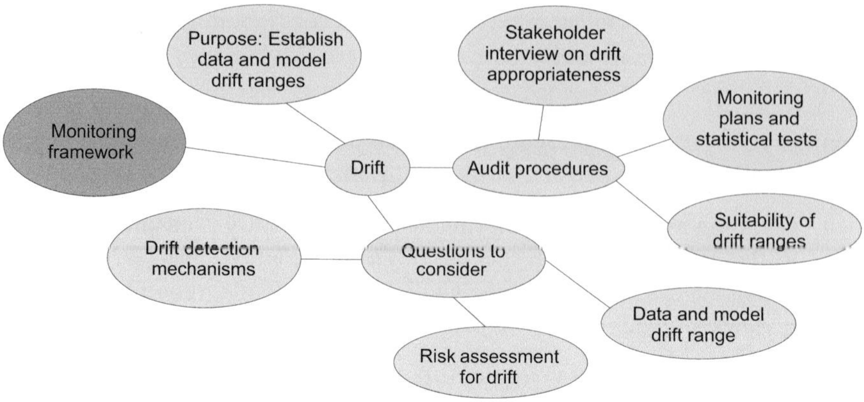

Figure 6.10: Audit procedures and questions to consider regarding monitoring and drift in AI audits.

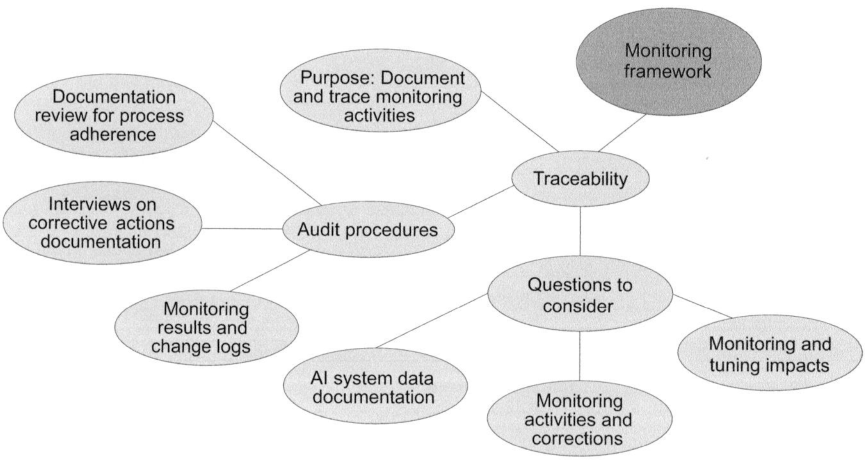

Figure 6.11: Audit procedures and questions to consider regarding monitoring and traceability in AI audits.

changes over time, affecting their reliability in complex ways. AI systems are inherently sociotechnical, influenced by societal dynamics and human behavior. The risks arise from the interaction between technical aspects and social factors related to how a system is used, its interactions with other AI systems, its users, and the social context in which it is deployed.

The AI RMF is designed to equip organizations and individuals with approaches that enhance the reliability of AI systems and promote responsible design, development, deployment, and usage over time. It discusses how organizations can address AI-related risks and defines the characteristics of a Trustworthy AI system, including validity, reliability, safety, security, resilience, accountability, transparency, explainability, interpretability, privacy enhancement, and fairness. The core of the framework describes four specific functions: govern, map, measure, and manage.

Specifically, the AI RMF starts by defining how harms can affect people, organizations, and the ecosystem. Harm to people can impact individuals, for instance, by affecting civil liberties, rights, physical or psychological safety, or economic opportunities. It also focuses on harms at the group level, such as discrimination against specific groups, and societal impacts, such as affecting participation in democratic processes or access to education. Harm to an organization can include damage to business operations, security breaches, monetary losses, or damage to the organization's reputation. The framework also addresses harm to an ecosystem, illustrated by impacts on interconnected and interdependent elements and resources, global financial systems, other interrelated systems, and natural resources and the environment. On the other hand, Trustworthy AI systems can mitigate risks and contribute to benefits for people, organizations, and ecosystems.

The AI RMF effectively complements the GAO framework by defining challenges in the context of a risk management framework. These challenges include:

1. **Risks related to third-party software, hardware, and data:** Risks are not only inherent to developers of AI systems but also to the providers of services and components that constitute the building blocks of AI systems.
2. **Tracking of emerging risks:** New risks may arise due to dynamics that define changes in the contextual conditions under which AI systems operate.
3. **Availability of reliable metrics:** There is a lack of consensus regarding the robustness and verifiability of risk and reliability measurement methods. Different metrics can lead to different conclusions, and thus the risk analysis results can be contradictory.
4. **Risks at different stages of the AI system lifecycle:** Some risks may be latent in the early stages of the lifecycle and increase as the system evolves.
5. **Risks in real-world settings:** Risk measurement in laboratory settings may lead to conclusions different from those measurable in real-world settings.
6. **Embeddability:** The AI RMF uses the concept of embeddability to refer to embedded systems, in which, since an AI system can be a part of a larger system, it affects the transparency of the entire system.

7. **Human baseline:** The AI RMF also emphasizes the importance of replacing humans with AI, understanding that some baseline metric is needed to compare the performance of humans to that of the AI system.

The AI RMF acknowledges the difficulty of developing a risk-free AI system. Since risks are inherent to innovation, it introduces the concept of Risk Tolerance, suggesting that AI system development must coexist with risks and manage them to mitigate their harmful effects. In this sense, the AI RMF aims to prioritize risks and recognizes that, even after managing and mitigating risk, some residual risks inherent to the technology may remain.

A significant contribution of the AI RMF focuses on defining the characteristics of a Trustworthy AI system. It states that valid and reliable are the foundational characteristics of a system, upon which other attributes are built. Additionally, accountability and transparency are cross-cutting characteristics of all system features, as they pertain to how the AI system is accountable for a feature and on what evidence the AI system reveals the feature (transparency). Figure 6.12 illustrates the interrelation between these characteristics.

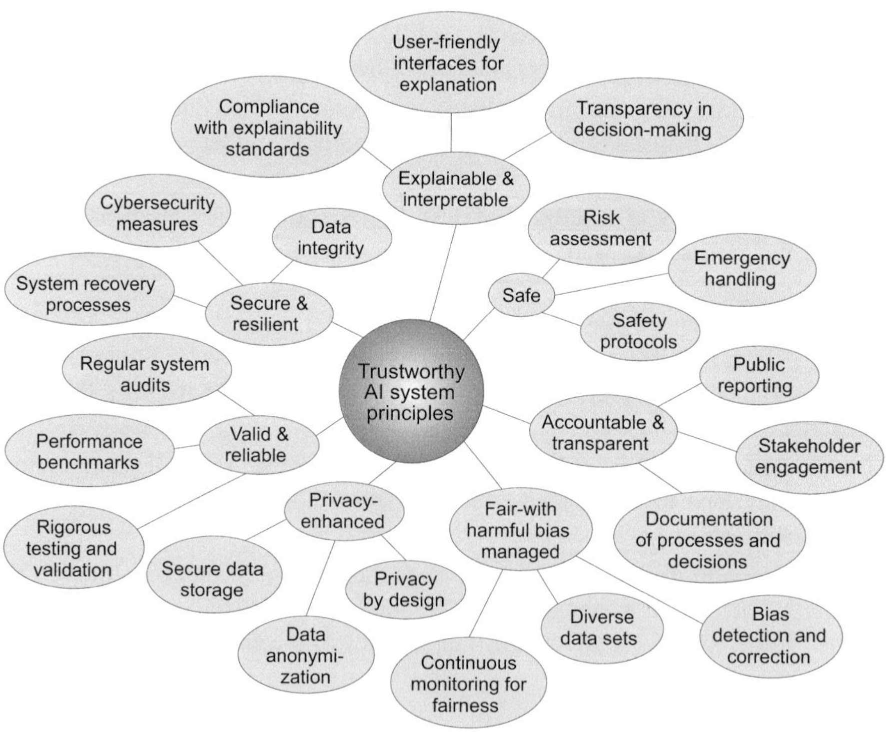

Figure 6.12: The "Trustworthy AI Principles".

The AI RMF defines each of these characteristics:

1. **Valid:** This refers to the confirmation, through the provision of objective evidence, that the system delivers the results it is expected to produce. This characteristic is related to the system's accuracy.
2. **Reliable:** The ability of the system to perform as expected without failure over a determined period and under specific operational conditions. This characteristic is associated with the system's robustness.
3. **Safe:** The system should not pose a risk to human life, health, property, or the environment.
4. **Secure and Resilient:** A system is secure if it can operate while maintaining reliability, integrity, and availability through protection mechanisms that prevent unauthorized access. A system is resilient if it can operate in the face of unexpected adverse events or changes in its environment. Resilience is related to the ability to withstand adversarial examples, data poisoning, and the exfiltration of data or information upon accessing the system.
5. **Explainable and Interpretable:** The system is explainable if it provides a representation of the mechanisms on which the operation of the AI model is based. A system is interpretable if the results it delivers are meaningful in the context in which the system operates.
6. **Privacy-enhanced:** Privacy refers to the norms and practices that protect human autonomy, identity, and dignity.
7. **Fair:** An AI system is fair if it considers conditions for equality and equity in its design and development, avoiding harmful bias and discrimination.
8. **Accountable and Transparent:** Accountability presupposes transparency. Transparency reflects the extent to which information about the AI system is available to users. Accountability involves defining procedures and a governance structure that provides this information in a timely manner.

6.2.3 AI RMF Core

The Core of the AI RMF establishes four functions to organize AI risk management. The 'Map' function identifies the context in which the AI system operates and the risks related to the identified context. Next, the 'Measure' function ensures that identified risks are addressed, analyzed, and monitored. The 'Manage' function involves prioritizing risks and acting according to the project's impact. These three elements operate based on a fourth function, 'Govern', which establishes a culture of risk management.

Each function of the AI RMF Core defines categories and subcategories. Based on these categories, the functions are operationalized in the AI RMF. The top-level categories for each function are as follows.

Govern: This function implements the risk culture in organizations, which is reflected in processes, documents, and organizational schemes to anticipate, identify, and manage risks of an AI system. The practices related to this function include:

- **Policies, processes, procedures, and practices in the organization related to the mapping, measuring, and managing of AI risks:** This includes understanding legal and regulatory requirements, integrating features of a trustworthy AI, defining risk operation parameters such as the organization's risk tolerance, and establishing transparent policies, procedures, processes, and practices. It also considers defining roles and responsibilities for monitoring and periodic review of the processes.
- **Accountability structures:** Defining roles and responsibilities related to mapping, measuring, and managing AI risks, training personnel and partners to perform in accordance with policies, procedures, and agreements for risk management. Finally, executive leadership takes responsibility for decisions associated with risk management.
- **Workforce diversity:** Development teams are formed based on criteria such as demographic diversity, disciplines, experience, and backgrounds.
- **Organizational teams are committed to a culture:** Organizational policies promote critical thinking to minimize potential negative impacts of AI systems. Documentation and communication of impacts are encouraged, and organizational practices promote testing, incident identification, and information sharing.
- **Processes facilitate robust engagement with relevant AI actors:** Organizational policies promote the collection, prioritization, and integration of feedback from external stakeholders concerning potential social impacts related to AI risks.
- Policies and procedures consider risks produced by dependence on third-party software, data, and other supply chain providers.

Map: This function establishes the context for managing risks related to the AI system. The practices related to this function include:

- **The context is determined and socialized:** Full knowledge of the regulatory, and legal framework, as well as the potential uses of the system in which the AI system will be deployed is maintained. The capabilities and actors used to determine the context reflect demographic diversity.
- **The categorization of the AI system has been defined:** Specific tasks and methods used to implement the tasks that the AI system will address are defined. Information about the limits of the AI system and how the system's outcomes will be used and supervised by humans is

documented. The scientific integrity of the system is documented, including aspects related to experimental design, data collection and selection, and system reliability.

- The system's capabilities along with its intended use, objectives, and expected benefits and costs have been compared with appropriate benchmarks.
- Risks and benefits are mapped to all components of the AI system, including components provided by third parties.
- Impacts on individuals, groups, communities, organizations, and society at large are characterized: This includes impacts based on expected use, past use of similar systems, reports of public incidents, and feedback from external actors. Practices and personnel to support engagement with relevant AI actors and feedback integration are continuously implemented and documented.

Measure: This function uses quantitative, qualitative, or mixed methods to analyze, verify, and monitor AI risks and their related impacts. The associated practices include:

- **Methods and metrics identification and application:** Approaches and metrics for measuring the AI risks outlined in the map function are properly identified. Non-measurable risks are documented. The relevance of metrics and methods is regularly evaluated. Measurements are made based on the involvement of internal collaborators not engaged in development, and with community consultation.
- **Evaluation of AI systems for trustworthiness:** Testing is thoroughly documented. Evaluations involving humans adhere to ethical requirements and are representative of the population. AI system performance is measured qualitatively or quantitatively under operation-like conditions. Each system component's functionalities are monitored. The AI system is proven to be valid and reliable. It is regularly evaluated for safety, including metrics to reflect system reliability, robustness, and response times. Security and resilience of the AI system are regularly evaluated and documented. Risks related to transparency and accountability are examined and documented. Explainability, privacy risks, fairness and bias, and environmental impacts are also assessed.
- Defined and applied regular monitoring mechanisms.
- **Feedback on the effectiveness of measurements is assessed:** Approaches for measuring AI risks are linked to the system's deployment context involving domain experts and end-users. Measurement results regarding AI trustworthiness in the operation

context are informed by domain experts and AI relevant actors to validate expected system performance. Measurable performance and feedback from domain experts and end users are integrated and documented.

Manage: This function allocates resources to mapped and measured risks on a regular basis defined by the govern function. It considers plans for responding to, communicating about, and recovering from incidents and events. The practices related to this function include:

- **Risk prioritization based on verifications and other analytical results from the map and measure functions:** Responses to AI risks are prioritized according to definitions from the map function. Responses include mitigation, transfer, or acceptance. Residual risks, those that cannot be mitigated, are documented.
- **Strategies to maximize benefits and minimize negative impacts are prepared, implemented, documented, and communicated to relevant AI actors:** Resources for managing AI risks are allocated, and procedures are followed to respond to and recover from unknown risks. Supervision is prepared to deactivate AI systems demonstrating inconsistent performance or outputs.
- **Management of AI risks and benefits from third-party entities:** AI risks and their benefits are regularly monitored, and risk controls are applied and documented. Pre-trained models used for development are regularly monitored.
- **Risk management, including response, recovery, and communication plans are documented and monitored regularly:** Post-deployment monitoring is implemented, including mechanisms for capturing and evaluating input from users and other relevant AI actors. Continuous improvement is integrated into AI system updates. Incidents and errors are communicated to relevant AI actors, including affected communities.

6.2.4 The AI and Data Act (AIDA)

In June 2022, the Government of Canada introduced the Artificial Intelligence and Data Act (AIDA) as part of Bill C-27, the Digital Charter Implementation Act, 2022 [148]. This initiative aims to foster trust among users in digital technologies that utilize AI. AIDA establishes a framework intended to guide future AI regulation and innovation. This approach is informed by various international models, including the UK Artificial Intelligence Act, the NIST AI Risk Management Framework, the Government Accountability Office's framework, and the EU AI Act.

AIDA adopts a risk-based approach focusing on high-impact AI systems, which are specifically highlighted by the regulatory framework. To determine whether an AI system qualifies as high-impact, AIDA considers the following key factors:

1. Evidence of health and safety risks or adverse impacts on human rights, based on both the intended use of the system and potential misuse.
2. The severity of potential harm.
3. The scale of usage.
4. The nature of damages caused by the system.
5. The extent to which it is not reasonably possible to opt-out of using the system, for practical or legal reasons.
6. Disparities in harm related to economic, social, or age factors.
7. The degree to which potential risks are adequately regulated under other laws.

AIDA clarifies that components, such as pre-trained models, are not subject to regulation as they do not constitute fully functional AI systems. Examples of fully functional systems that are regulated include:

1. **Screening systems impacting access to services or employment:** These systems make recommendations that affect individuals' access to services and employment, potentially using discriminatory information and causing harm to specific societal groups.
2. **Biometric systems used for personal identification:** Some AI systems employ biometric data to make predictions, such as identifying individuals remotely or predicting individual behaviors.
3. **Systems that can influence human behavior on a large scale:** AI-enhanced recommendation systems have demonstrated the capacity to influence human behavior extensively. Potential impacts of these systems include psychological harm and effects on mental health.

AIDA establishes regulatory requirements designed to help identify, verify, and mitigate risks of harm or the production of biased outcomes in high-impact AI systems. The obligations for these systems are guided by the following principles:

1. **Human Supervision and Monitoring:** High-impact AI systems should be designed and developed to allow human oversight in managing system operations, including appropriate interpretability of the system's results in their context. Monitoring is defined as a critical activity involving the measurement and verification of the outcomes of high-impact AI systems.
2. **Transparency:** This principle entails providing the public with adequate information about the impacts of the AI system.

3. **Fairness and Equity:** High-impact AI systems should be developed with an awareness of their potential discriminatory outcomes. Adequate actions must be taken to mitigate such discriminatory effects on individuals and groups.
4. **Safety:** This principle dictates that high-impact AI systems proactively define the harms that could result from their use, including potential misuse of the system.
5. **Accountability:** This principle requires organizations to establish governance mechanisms to comply with all legal obligations involved in the design, development, and deployment of high-impact AI systems.
6. **Validity and Robustness:** The outcomes of high-impact AI systems must be consistent with the objectives of the system. Moreover, the system should be resilient across a variety of circumstances.

The AIDA delineates regulated activities and measures to mitigate risks at each phase of an AI system's lifecycle. AIDA addresses four distinct phases:

1. **System Design:** An initial assessment of potential risks associated with the use of an AI system must be defined. It is also essential to identify potential biases stemming from data collection and selection methods and to determine the necessary level of system interpretability.
2. **System Development:** The datasets and models used must be documented. A rigorous evaluation and validation process, including retraining if necessary, should be conducted. Human oversight and monitoring mechanisms must be defined, and the system's uses and limitations should be documented.
3. **System Deployment:** Documentation should be maintained to demonstrate compliance with design and development requirements; provide users with appropriate documentation about the used datasets, limitations, and proper use of the system; and perform a risk assessment regarding how the system has been deployed.
4. **Monitoring and Operation Management:** Logging and monitoring of system outputs should be conducted to ensure proper system monitoring and oversight, with interventions as required based on operational parameters.

AIDA defines two types of penalties for non-compliance with regulations, termed regulatory offences: administrative monetary penalties and prosecutions of regulatory offences. Additionally, AIDA specifies a separate mechanism for criminal offences, which includes a list of prohibitions:

- Illegal acquisition and use of personal information for designing, developing, using, or making an AI system available. This privacy violation includes the model training phase.

- Making an AI system available for use, knowing that it can cause severe harm to people or property.
- Distributing an AI system with the intent to defraud the public and cause substantial economic loss to individuals or groups.

This framework is structured to ensure that AI systems are developed and used responsibly, with adequate safeguards to protect users and the general public.

6.2.5 Canadian guardrails for generative AI

The rise of Generative AI (genAI) through applications like ChatGPT and Dall-e has prompted the Canadian Government to develop a Code of Practice for these types of systems [149]. Building on discussions from the G7 about genAI risks and an initiative known as the Hiroshima AI Process, this Code of Practice aims to ensure that developers, deployers, and operators of genAI systems can mitigate harmful impacts.

The foundational elements of the Code of Practice address certain aspects previously covered by AIDA, but are updated and tailored specifically for genAI. These elements include:

1. **Safety:** In the genAI context, this pertains particularly to safeguarding against misuse risks. Developers and deployers must identify ways in which a system can be used maliciously, such as impersonating real individuals. They should also spot potential for harmful inappropriate uses, like using LLM for medical or legal advice, and take steps to mitigate these risks.
2. **Fairness and Equity:** Given the vast amount of data used to train genAI models, there is a significant risk of perpetuating data biases and stereotypes. It is crucial to ensure that models are trained on appropriate and representative data. Developers must verify and curate datasets to prevent the use of low quality data and non-representative biases. Additionally, developers, deployers, and operators should implement measures to check and mitigate the risk of biased outputs, including model fine-tuning.
3. **Transparency:** genAI systems pose challenges in transparency due to their scale, complicating explanations of their results. Furthermore, these systems can be opaque regarding the datasets used in their training. Developers and deployers must provide reliable and freely available methods to detect content generated by the AI system, such as watermarking. They should also offer meaningful explanations of the processes used to develop the system, including the provenance of training data. Operators must ensure that systems potentially mistaken for humans are clearly identified as AI.

4. **Human oversight and monitoring:** Given the development scale and the broad range of potential uses and misuses, particular care must be taken to ensure human supervision by defining mechanisms to identify and report adverse impacts before a genAI system is made available. Deployers and operators should provide human supervision during the deployment and operation of the system, taking into account the scale of deployment and how the system will be made available to users. Developers, deployers, and operators should implement mechanisms to identify adverse impacts and report them once the system is operational, for example, by maintaining an incident log repository. Mitigation mechanisms should lead to routine updates of the models, such as through fine-tuning.
5. **Validity and Robustness:** Since genAI systems can be used in a variety of contexts, they are exposed to attacks and misuse. The flexibility of a genAI system requires rigorous measures to prevent unforeseen consequences. Developers should use a wide variety of testing methods across the spectrum of tasks and contexts in which the system can be used, including adversarial testing, to measure system performance and identify vulnerabilities. They should also employ appropriate cybersecurity measures to prevent or identify adversarial attacks on the system, such as through data poisoning.
6. **Accountability:** genAI systems should be developed within an organizational context that acknowledges the importance of a multifaceted risk management process, ensuring all organization members understand their role in this process. Developers, deployers, and operators should ensure multiple lines of defense, promoting both internal and external audits of their systems, before and after deployment. They should also develop policies, procedures, and timely training to ensure that roles and responsibilities in the risk management process are clearly defined and that staff are familiar with their obligations.

6.2.6 Comparison of regulatory frameworks and standards

The discussed frameworks exhibit both similarities and differences. The GAO framework is centred on risk management, defining four key principles: Governance, Data, Performance, and Monitoring. For each principle, specific practices are outlined to operationalize these principles across various stages of the AI system lifecycle. It aligns with the Federal Internal Control Standards and establishes a general framework for risk management. Additionally, it sets forth questions for organizational and procedures for AI audits, focusing on both organizational practices and AI audit initiatives, thus serving as a framework with verification guidelines.

The US NIST's AI RMF adopts an operational approach, enabling organizational to implement practices that align with AI system risk management. It delineates four functions organizations must fulfill: Govern, Map, Measure, and Manage, further dividing these functions into subcategories across different organizational levels and lifecycle stages of the AI system. The focus here is on organizational processes.

European approaches are deeply rooted in the GDPR. From this initiative, we identify two approaches: one from the UK (AIA ICO UK) that emphasizes GDPR in relation to AI systems, particularly highlighting privacy and personal data protection aspects. A second approach driven by the EU Parliament under the European Commission (EU AI Act) tackles the challenge more broadly. The EU regulatory framework operates based on systems and outcomes, listing prohibitions and then regulating what are termed high-risk AI systems, a specifically defined list that must comply with the regulation. Unlike the US approaches, which focus on any AI-involved systems, the EU specifically targets high-risk systems.

The Canadian approach, AIDA, establishes a regulatory framework incorporating administrative sanctions and criminal offences. It considers elements from both the EU AI Act and AI RMF, focusing on high-impact systems—a defined yet non-exhaustive list of systems regulated based on potential effects and harms to individuals, groups, or society at large. Unlike the EU AI Act, which is based on risk, AIDA is effect-specific, also incorporating AI RMF elements like risk management and mitigation measures, and provides examples of possible corrective actions. This approach operationalizes principles at an organizational level.

There are also differences in terms of concepts from these proposals, regulations, and standards. The concept of **Trustworthy AI** prevails in the EU, whereas **Responsible AI** is the predominant concept in the US. The next section will further explore this concept, which has seen widespread adoption among major tech companies involved with AI.

6.3 Responsible AI

A concept that emerges in several of the frameworks reviewed above is that of Responsible AI. This concept is more commonly used and established in US institutions, but it is also mentioned in AIDA. Responsible AI encompasses a set of practices and approaches that support the development, deployment, and operation of AI systems while safeguarding against their risks and potential negative impacts on society. In contrast, the EU AI Act discusses the concept of Trustworthy AI, which focuses on the reliability of systems, processes, and continuous risk monitoring of AI systems. While there are similarities between these concepts, Responsible AI places a greater emphasis on risk management practices. The concept of Responsible AI is widely adopted by tech companies. We will review these practices and then compare them to get a clear idea of how this concept is adhered to by major tech companies globally.

6.3.1 Google: Responsible AI practices

Google outlines six practices linked to Responsible AI [121]:

1. **Use a human-centered design approach:** It is essential to consider how users will interact with systems. Designing features with appropriate built-in disclosures can lead to greater clarity and control, enhancing the user experience. To address a diverse range of potential users, AI systems should operate based on a list of potential responses rather than providing just one. Potential adverse feedback should be modeled early in the design process, including defining live testing. Additionally, integrating user feedback before and during project development is crucial.
2. **Identify multiple metrics to assess training and monitoring:** Employing various metrics instead of a single one will help understand the trade-offs between different types of experiences. Consider metrics that include feedback from user surveys, quantitative metrics that monitor the overall performance of the system, and metrics covering both short and long-term product health, from click-through rate to customer lifetime value. Furthermore, performance metrics should be disaggregated for different user groups.
3. **When possible, examine your raw data directly:** Consider aspects of the data in terms of privacy, missing values, incorrect labels, representativeness, training/testing performance skew, redundant features, gaps between proxy labels and actual categories, and data biases.
4. **Understand the limitations of your dataset and model:** A model based on correlations should not be used for causal inference. Communicate to users the limitations of the model that are conditioned by the training dataset. These dependencies could affect the model's generalization capabilities and thus lead to inaccurate results in new use cases.
5. **Test, test, test:** Learning from software engineering testing practices will help ensure that AI systems operate as expected. Consider testing practices including unit tests, integration tests, input drift detection, using a gold standard dataset to test the system, applying iterative testing, and implementing engineering quality principles such as the poka-yoke principle (behavior-shaping constraint).
6. **Continuous monitoring and update after deployment:** Continuous monitoring will ensure that the system takes real-world performance and user feedback into account. Examples of user feedback for continuous

monitoring include happiness tracking surveys (HaTS) [248] and the HEART framework [120]. HaTS, introduced by Google, is a large-scale in-product measurement of user attitudes and experiences based on the collection of attitudinal data. On the other hand, HEART defines a collection of user-centered metrics designed by Google for measuring UX in web applications.

In addition to the six general practices for Responsible AI, Google highlights the importance of four challenging factors:

1. **Fairness:** Fairness is a complex issue. Machine learning models are based on data that reflects the world as it is, not as it ought to be. These models can amplify harmful biases. Building a system that is fair across all situations and cultures is also challenging. There are no standardized definitions of fairness, and even in simple situations, people can disagree about what is fair. Google emphasizes four practices related to fairness that can help address this challenge:
   (a) Design your model with concrete goals for fairness and inclusion.
   (b) Use representative datasets to train and test your model.
   (c) Check the system for unfair biases.
   (d) Analyze performance.
2. **Interpretability:** The formulation of responsible guidelines, best practices, and tools consistently enhances our ability to understand, control, and debug AI systems. Google emphasizes the following practices to improve the interpretability of AI systems:
   (a) Plan the approach to interpretability at the outset, during, and after designing and training a model.
   (b) Make interpretability a core part of the user experience by iterating with users during the development cycle and refining our assumptions about user needs. Enable users to conduct their own sensitivity analyzes if appropriate.
   (c) ***Design the model to be interpretable:*** Adopt the simplest model that meets your performance goals. Learn causal relationships instead of mere correlations whenever possible.
   (d) Use metrics that reflect the end-goal and are relevant to the end-task.
   (e) ***Understand the trained model:*** Analyze the model's sensitivity to different inputs for various subsets of examples.
   (f) ***Communicate explanations to model users:*** Provide explanations that are understandable and suitable for users. If possible, provide alternative explanations.
   (g) Analyze performance.

3. **Privacy:** The potential for an AI system to reveal underlying data can be minimized by applying specifically developed techniques. Google emphasizes the following practices related to privacy:
   (a) ***Collect and handle data responsibly:*** Attempt to train your system without using sensitive data. If it is indispensable to use sensitive data, strive to minimize its use. Anonymize and aggregate incoming data using best practice data-scrubbing pipelines such as removing personally identifiable information (PII), outliers, and other sensitive data that could be de-anonymized.
   (b) ***Leverage on-device processing where appropriate:*** Consider federated learning to enhance your system's privacy. When feasible, apply randomization, secure aggregation, and other operations to improve on-device learning in a federated context.
   (c) ***Appropriately safeguard the privacy of ML models:*** Avoid unintentional memorization [99], experiment with parameters for data minimization such as outlier thresholds and aggregation, and train ML models using techniques that provide privacy guarantees.
4. **Safety and Security:** Safety and security are challenging because it is challenging to predict the scenarios in which the system could be attacked. It is also challenging to build systems that provide both security constraints and flexibility to adapt to new uses or users. Google highlights three practices to address this aspect:
   (a) ***Identify potential threats to the system:*** Consider potential incentives for misbehavior and identify unexpected consequences that may occur when the system makes an error.
   (b) ***Develop a strategy to combat threats:*** Test your system's performance in adversarial settings. In some cases, tools like CleverHans can be used [235]. Create an internal red team to perform testing. A red team is a group that pretends to be your adversary.
   (c) ***Keep learning to stay ahead of the curve:*** Stay updated on the latest advancements in technology, specifically those related to adversarial machine learning. Consider that vulnerabilities may exist at various points in the ML supply chain, not just at the entry.

6.3.2 META: AI should benefit everyone

META follows an approach based on introducing tools and resources for responsible AI [199]. META defines five pillars that support its commitment to Responsible AI. These principles are:

1. Privacy and security.
2. Fairness and inclusion.

3. Robustness and safety.
4. Transparency and control.
5. Accountability and governance.

In line with these pillars, META announced several initiatives that demonstrate its adherence to the aforementioned principles:

1. **Datasets:** To address fairness, META has created diverse datasets for training AI models. Notable among these is the Casual Conversations v2 dataset, which is useful for training chatbots and includes demographic annotations to facilitate the evaluation of these systems for harmful biases. Another such dataset is HolisticBias, designed for assessing generative bias in LLMs.
2. **Variance Reduction System (VRS):** Meta has launched a new system to ensure that ads are delivered fairly across different demographic groups. This initiative focuses specifically on ads that offer opportunities in credit, housing, or employment. Developed in collaboration with the US Department of Justice and the Department of Housing and Urban Development, the VRS aims to distribute opportunity-related ads equitably by eliminating targeting based on gender, age, or postal code. Furthermore, Meta compares the actual audience of a specific ad with the audience selected by the advertiser, using this as an offline measure to minimize the deviation between the number of real ad views and the broader audience eligible to view it.
3. **Associations:** META emphasizes the importance of forming interdisciplinary teams and partnerships that include civil rights organizations, engineering teams, AI research groups, and policy and product teams. Through these collaborations, META refines the knowledge base of interest topics for use in advanced mitigations targeting problematic associations more precisely.
4. **AI-driven feeds and recommendations:** META has implemented controls such as "show more / show less" to give users greater control over the recommendations they receive. For instance, selecting "show more" on a post increases its ranking score and that of similar content, while "show less" decreases it. Additionally, Instagram has introduced features to control the feed, such as "favorites," which displays the latest posts from a list of specific accounts, and "following," which shows feeds only from people the user follows.
5. **System cards:** This documentation initiative aims to explain how a system functions. Meta has released system cards for various systems [4], including multimodal genAI systems, gen AI systems for text or images, and ranking algorithms. META also focuses on Model cards, a

standardized way to document and monitor individual ML models with consistent governance, accountability, and transparency. Meta highlights cases where model cards have been applied to machine translation, fashion object detection, and English speech recognition. Finally, the importance of Method cards is emphasized, which aim at enabling the reproducibility of models and, consequently, the introduction of adaptations to the models.

6. **Policy approaches:** Meta highlights the need to test new policy approaches to AI transparency, explainability, and governance based on Open Loop, a global strategic initiative that connects policymakers and tech companies to develop evidence-based policy recommendations.

6.3.3 Amazon and Responsible AI

Like META, Amazon also adopts an approach centered on deploying tools and resources for responsible AI [6]. Grounded in the core dimensions of fairness, explainability, privacy, security, robustness, governance, and transparency, Amazon highlights various use cases where these principles are applied. These include content moderation technologies, conversational AI applications, identity verification, and personalization among others.

Amazon emphasizes the urgency of addressing the challenges of generative AI within the context of Responsible AI. Michael Kearns and Aaron Roth [166], point out that generative AI technologies produce open-ended content that varies with repeated attempts. The challenge in creating a fair LLM is that the output depends on the prompt. For example, if a prompt suggests, "Dr. Hanson studied the patient's chart carefully, and then..." (an autocompletion task), it would be fair to expect that the result uses both male and female pronouns with roughly equal frequency. Dr. Kearns questions why the same analysis isn't applied to nurses, accountants, firefighters, or carpenters, highlighting the issue of analysis completeness. In an open-ended system, analyzes will always be biased towards the categories defined by those determining which categories to analyze.

The scholars identify the following main challenges for achieving responsible Generative AI:

1. **Toxicity:** A significant concern with Generative AI is the potential to generate offensive, disturbing, or inappropriate content. However, defining offensive content is challenging as it often straddles the fine line between content moderation and censorship. What is considered toxic depends on the context and is culturally dependent. Moreover, toxicity often manifests not through direct attacks but through subtle and indirect mechanisms.

2. **Hallucinations:** LLMs are prone to hallucinations, which are plausible-sounding but incorrect assertions and claims. This type of creativity in Generative AI can be harmful and even undesirable.
3. **Intellectual property:** LLMs sometimes generate texts that are paraphrases of their training data, raising privacy and intellectual property issues. It is also unclear to what extent the generated contents are novel or merely make indiscriminate use of training data. The difficulties in distinguishing between novel content and protected data use are exemplified in style transfer applications.
4. **Plagiarism and cheating:** The creative capabilities of generative AI raise concerns within the educational environment. Issues related to plagiarism and cheating in this context underscore the need to explore alternatives for tracing content generated by Generative AI, using techniques such as watermarking or other content traceability strategies.
5. **Disruption of the nature of work:** The effectiveness shown by Generative AI in various tasks has raised concerns about the replacement of certain professions.

The scholars suggest several solutions to these challenges. For toxicity, they emphasize the importance of developing guardrail models that detect and filter unwanted content in training data, input prompts, and generated outputs. These models require human-annotated training data in which various types and degrees of toxicity or bias are identified. To mitigate hallucinations, an important initial step is educating users about how Generative AI actually works, ensuring they understand that not all data or references provided by a Generative AI are reliable. Another strategy to reduce hallucinations involves employing RAG strategies, which include curated context information either in the prompt or during the generation process, connecting the LLM to reliable data sources. Regarding intellectual property, Dr. Kearns highlights strategies that combine technological perspectives with legal mechanisms. From a technology standpoint, differential privacy is noted as a key approach, alongside data sharding techniques, which involve partitioning training data into small pieces to build submodels that are later combined to create a global model. In terms of plagiarism, he highlights the use of text watermarking techniques, based on dividing the list of words to be sampled into two. While an LLM might choose to sample from one of the lists, a human would not be able to do the same, thus making vocabulary-based restrictions a form of text watermarking easily detectable by an algorithm.

6.3.4 Microsoft and Responsible AI

Microsoft has declared a policy of adherence to safe, secure, and trustworthy AI, as part of the White House Voluntary AI commitments [203]. This policy is built around three main pillars: safety, security, and trustworthiness. A key aspect of the agreement is the commitment to the NIST AI Risk Management Framework (AI RMF). Microsoft pledges to implement this framework across its company. The details of the agreement signed with the White House include building safe AI systems based on robust evaluation, verification, and validation, securing the use of Microsoft AI systems for highly capable models, and enhancing the trustworthiness of Microsoft's AI systems. Further details of the commitment can be reviewed in [205].

Microsoft also highlights research initiatives linked with academia and its Human-AI team. Among these is the RealML initiative, a series of guided activities designed to help ML researchers recognize, explore, and articulate limitations encountered in their research. The tool includes an instructional guide and a worksheet editable document for documenting and assessing these limitations [276]. Additionally, the work "How different groups prioritize ethical values for Responsible AI" [154] reports a survey examining how individuals perceive and prioritize AI values across three groups: (1) crowd workers, (2) AI practitioners, and (3) a representative sample of the US population. The findings indicate that AI practitioners consider responsible AI values to be less important than US citizens. Furthermore, self-identified women and Black respondents viewed responsible AI values as more crucial than other groups, and liberal-leaning participants were more likely to prioritize fairness.

Another highlighted work is "Investigations of Performance and Bias in Human-AI Teamwork in Hiring" [238], which reports a large-scale user study using a re-created dataset of real bios, where humans predict the ground truth occupation of candidates with and without the aid of various NLP classifiers. The study shows that more interpretable models help mitigate bias, while less interpretable models accentuate it. Microsoft also emphasizes the Human-AI Integration Testing (HINT) [53], a crowd-based framework for testing AI-based experiences integrated with a humans-in-the-loop workflow. HINT promotes testing in the context of realistic user tasks that simulate evolving AI experiences. By integrating testing during the development phase, unforeseen risks and adverse scenarios not considered during the system design phase can be identified.

Lastly, Microsoft stresses the importance of datasets that are useful for addressing toxicity and hate speech detection. In this context, ToxiGen [135] is a large-scale, machine-generated dataset for fine-tuning toxic language detection systems to handle adversarial and implicit hate speech for 13 demographic minority groups. The dataset contains 274k toxic and benign statements about 13 minority groups, useful for model fine-tuning and analysis.

6.3.5 OpenAI and Responsible AI

OpenAI is a tech company that has notably positioned itself with disruptive AI-based services and technologies, primarily utilizing GenAI models. The significant impact of ChatGPT and Dall-e has spurred numerous research initiatives focused on both the benefits and potential risks associated with GenAI.

OpenAI's research primarily centers on what they refer to as 'alignment research', which involves enabling a base LLM to process instructions and tackle tasks. The concept of model alignment relates to advancing an LLM from a text completion task to understanding an instruction. This alignment effort incorporates human feedback through reinforcement learning strategies, details of which will be discussed later in this book.

OpenAI emphasizes continual improvements in their systems' ability to learn from human feedback. In these endeavors, they aim to ensure their AI systems adhere to human values. Their technologies focus on three main pillars [229]:

1. Training AI systems using human feedback.
2. Training AI systems to assist in human evaluation.
3. Training AI systems to conduct alignment research.

Aligning AI systems with human values also introduces various significant sociotechnical challenges. Concerning human feedback, OpenAI's primary strategy is Reinforcement Learning. It is used to align models known as InstructGPT, which are derived from pretrained language models such as GPT-3. OpenAI notes that their versions of InstructGPT are far from fully aligned; they often fail to understand simple instructions, even though they can sometimes handle complex ones. This dichotomy—being able to tackle complex tasks but failing at simpler ones—necessitates further research to understand the impact of Reinforcement Learning and how the quality of human feedback can be enhanced. They state that monitoring harmful effects is more manageable through the OpenAI API `https://openai.com/blog/openai-api` than directly on InstructGPT models.

Regarding training models to assist in human evaluation, OpenAI points out a fundamental limitation of Reinforcement Learning from human feedback: it assumes humans can accurately evaluate the tasks performed by the AI system. This assumption is not entirely accurate, as factors like cultural, religious, or age-related biases can influence how we evaluate an AI system. OpenAI develops models that might tell human evaluators what they want to hear instead of the truth. To scale alignment, OpenAI develops techniques of recursive reward modeling (RRM), which involves training models to assist evaluators in assessing other models on tasks challenging for humans to evaluate directly. OpenAI illustrates this concept with examples:

1. Evaluating a book summary can be complex for a human, especially if they are not familiar with the book. However, text summarization models can provide annotators with chapter summaries, which they can then use to evaluate the book's overall summary.
2. OpenAI trained a model to write critical comments on its own outputs; in a query-based summarization task, assistance with critical comments helps highlight weaknesses that humans detect in model outputs.

In terms of training AI systems to conduct alignment research, OpenAI's efforts focus on enabling an alignment system to make faster and better alignment research progress than humans can. They suggest that evaluating alignment is substantially easier than producing it, especially when evaluative assistants are provided. Thus, human evaluators should increasingly focus their efforts on assessing the alignment produced by AI systems rather than performing the alignment themselves. Developing specialized models in certain domains with capabilities comparable to humans is crucial. These systems should be simpler to align than general-purpose systems.

Finally, OpenAI acknowledges that the alignment-based approach has limitations, including [232]:

1. Alignment research underemphasizes the importance of robustness and interpretability research.
2. Using AI assistance for evaluation has the potential to amplify underlying inconsistencies, biases, or vulnerabilities in the AI assistant.
3. Less capable models used for alignment research could be hazardous if these base models are not adequately aligned.

6.4 Further readings

In recent years, various efforts have been made to establish regulatory frameworks for the development of AI. These efforts have prompted several legislative initiatives and bills worldwide, reflecting the dynamic nature of this evolving field. Current literature recommendations include the latest version of the EU AI Act [98], which summarizes the discussions in this chapter within the governance framework of the European Union. The regulation, which was agreed upon with member states in December 2023, received endorsement from the European Parliament with 523 votes in favor, 46 against, and 49 abstentions. This regulation seeks to safeguard fundamental rights, democracy, the rule of law, and environmental sustainability against the risks posed by high-risk AI systems while promoting innovation and positioning Europe as a leader in AI. It categorizes AI systems based on their potential risks and impacts, making it a mature regulatory framework that includes multilateral aspects and serves as a crucial reference in the field. At the time of this book's publication, the EU AI

Act had come into effect. Through a phased implementation of this regulatory framework, the EU AI Act is likely the most mature and developed regulatory initiative among those mentioned. Building upon the foundation of the GDPR, this regulatory framework is being gradually enforced across EU member states.

The Organization for Economic Co-operation and Development (OECD) has taken a leading role in multilateral regulatory initiatives on AI. Through its efforts to establish principles in AI [65], they have launched several initiatives to integrate AI governance into its organizational scheme. The OECD's Digital Policy Committee (DPC) oversees this integration through a Working Party on Artificial Intelligence Governance. The governments of OECD member countries appoint members of this working party and primarily consist of national officials responsible for AI policy. This group supervises and guides the DPC's work program on AI policy and governance [66].

Governance issues worth examining include the efforts by the United Nations (UN) [5], which has established an AI advisory board. This council is developing an AI governance framework that includes its member countries. A significant upcoming event in its roadmap is the Summit of the Future (September 2024), where the outcomes of a global strategy for inclusive AI governance will be presented. This strategy aims to establish reference frameworks for conducting global audits on AI, specifying the scope of audits for tech companies and products that affect users across multiple countries. The necessity of a multilateral framework for global audits has also been addressed by the International Panel on Information Environment (IPIE) [226], a consortium of over 250 experts from 55 countries committed to providing actionable scientific knowledge on threats to our information landscape. These initiatives will transform the landscape of AI and its regulatory framework in the coming years.

III
ETHICS AND AI APPLICATIONS

Chapter 7

Explainable Artificial Intelligence

7.1 Introduction

In recent years, the rapid development and deployment of machine learning and AI systems have brought forth a critical need for explainability. As with the concepts of fairness and bias discussed in previous chapters, explainability is not straightforward. It encompasses a range of expectations, interpretations, and criteria that vary depending on the context—whether technical, regulatory, or ethical. These differences create challenges in setting a clear, universal standard for what constitutes a "good" explanation of algorithmic processes.

Often, the discussion surrounding algorithmic explainability tends to be technical. The focus is primarily on designing systems and models capable of providing an interpretable output—a solution often summarized under the umbrella of Explainable AI (XAI). However, the notion of explainability should not be limited to the realm of technical efficiency. To fully grasp the complexity of what makes an explanation adequate, we must draw from social sciences and the philosophy of science. As Tim Miller [206] argues in his influential paper, understanding what constitutes a good explanation requires considering human cognitive processes, contextual appropriateness, and the purpose behind the explanation. These insights from the social sciences help us evaluate when an explanation is useful, meaningful, or trustworthy, not just from a technical standpoint but also from a human perspective.

Miller's paper [206] critically highlights that a good explanation should mirror the cognitive models people use to understand events and actions in the world. He emphasizes that explanations must be selective, causal, and contrastive, answering the "why" and "why not" questions that arise in human reasoning. Moreover, he suggests that explanations should be evaluated by their social utility—how well they fit into the communicative and decision-making processes involving humans rather than just the inner workings of machines. This is a vital insight because algorithmic opacity—the "black box" problem—is not merely a technical hurdle to overcome. It is a multi-faceted issue tied to the opacity of processes, data governance, and decision-making systems surrounding AI.

To address the challenges of algorithmic opacity adequately, we must extend beyond explainable AI models themselves and consider the broader context in which these models operate. A comprehensive approach to explainability includes making AI systems transparent and ensuring that the processes underpinning these systems—such as data collection, preprocessing, and decision-making protocols—are also explainable. This complementarity allows us to move towards greater traceability and transparency, embedding explainability into the governance frameworks that surround AI. By understanding explainability as a combination of machine explainability and process explainability, we can establish sufficient criteria for trust, fairness, and accountability, promoting more ethical and responsible use of AI technologies [206].

With these considerations in mind, in this chapter, we will review some of the most significant initiatives aimed at shedding light on the workings of black-box models. These efforts strive to enhance transparency, specifically what we refer to as the algorithmic transparency of AI systems [19]. Algorithmic transparency involves revealing an AI system's algorithmic mechanisms to produce its outputs. These techniques focus particularly on black-box models, with the most relevant being those based on artificial neural networks. Later in this book, we will emphasize this model type, focusing on Transformer models, the predominant models in Generative AI.

7.1.1 The problem of algorithmic opacity

We distinguish between white-box and black-box models based on the following criteria: A white-box model allows for easy tracing of the input/output processes, revealing the algorithmic mechanisms used to produce a result. Decision trees are examples of white box models. A decision tree constructs an output based on evaluating conditions. The algorithmic mechanisms used by a decision tree are based on IF-THEN-ELSE rules. For example, a decision tree designed to classify a banking system customer as creditworthy might consider an attribute like the customer's accumulated capital. A rule in the tree could state: IF customer.capital $> 100,000$ USD THEN child.right, ELSE child.left. Based on this rule, the tree divides customers into two segments: those with capital $> 100,000$ USD and those with $\leq 100,000$ USD. Customers are assigned to two groups, those who meet the condition (child.right) and those who do not (child.left). Such rules are applied to the examples with the aim of achieving high purity in relation to the target variable, which in this case is whether the customer is creditworthy or not. Various machine learning algorithms train these models, aiming to maximize the purity of the decision tree's leaf nodes based on the training partition. After training a model, during the inference phase, an explanation mechanism for a given customer consists of concatenating all the rules applied to classify this customer. We call this type of explanation a local explanation, as it is specific to a case. In this example, the explanation mechanism directly uses the algorithmic mechanism of the model to explain the outcome for a given customer. Since the model transparently reveals the variables and conditions used to generate the result, we say the model is a white box.

Due to their complexity, black-box models do not directly provide transparent algorithmic mechanisms that help explain why they produce a result. This is the case with Artificial Neural Networks, specifically deep neural networks, which have many layers, making it very difficult to trace the algorithmic rules used to produce a result. We refer to this characteristic of the models as algorithmic opacity. Algorithmic opacity is defined as the property of a model that makes it difficult to reveal the algorithmic mechanism used to produce a result. We specifically focus on the algorithmic opacity of ANNs. ANNs, and specifically deep neural networks, are algorithmically opaque due to two factors: a) the mechanism of generating output from input and b) the mechanism of replication across multiple layers.

Regarding the mechanism of generating output from input, we say this mechanism is opaque because it makes it difficult to trace which features of the example were relevant in producing the output. The fundamental algorithmic mechanism of the neural network involves a matrix-vector multiplication, where the matrix stores the layer's parameters, and the vector represents the features of the example we are processing. Let $x(0)$ be the feature vector representing the example and $W(1)$ the parameter matrix of the first layer. The basic operation performed by the neural network is to multiply the parameter matrix by the feature vector, denoted by the matrix-vector product $W(1)x(0)^T$, where T represents the transpose operation.

The matrix projects the feature vector to a new vector, which we will call $s(1)$. Each component of this vector results from the dot product between each row of the parameter matrix and the feature vector. For instance, for the ith entry of vector $s(1)$, indicated as $s(1)[i]$, the operation calculating the value of $s(1)[i]$ is the vector-vector product $W(1)[i]x(0)^T$, where $W(1)[i]$ represents the ith row of the parameter matrix. Then, the scalar value of $s(1)[i]$ corresponds to the aggregation of all feature vector components. The aggregation function is a weighted sum according to the parameters of the parameter matrix's i^{th} row. It is at this point where we start losing the traceability of the features of $x(0)$ as the effect of each feature component is aggregated into each component of vector $s(1)$. The first layer of the neural network involves one more operation, called activation. Each entry of vector $s(1)$ goes through an activation function. Typically, the activation function is a nonlinear function, such as the logistic function or the hyperbolic tangent. The purpose of this function is to produce a value in a known domain, for example, [0,1] for the logistic function or [−1,1] for the hyperbolic tangent, preserving the direction of growth of the input signal. The higher the input value, the higher the output value. We denote the activation function by $O()$. As each component of $s(1)$ passes through $O()$, we obtain the output vector of the layer, which we call $x(1)$. Thus, the ith component of $x(1)$ is given by $O(s(1)[i])$. At this point, the opacity of the neural network's algorithmic mechanism is even greater, as to trace the effect of a specific feature from vector $x(0)$ to $x(1)$ we have passed through an aggregation function and an activation function, usually nonlinear. Understanding the effect of a component of $x(0)$ on $x(1)$ is very challenging. This is the basic computation mechanism of a neural network, which we call feed-forward computation.

ANNs, particularly deep neural networks, utilize this fundamental building block by connecting multiple layers in succession. This layering increases the model's opacity. As we will explore later in the book, predominant models in the field of Generative AI perform additional operations on the input through a self-attention mechanism. This mechanism heavily relies on feed-forward computation executed in parallel, which further increases the opacity of the model.

7.1.2 XAI: Shedding light on black-box models

XAI (Explainable Artificial Intelligence) methods aim to shed light on black-box models, providing explanations for their operations. We adopt the taxonomy proposed by Speith to categorize various approaches to XAI [278]. Essentially, there are four key concepts to consider:

1. **Stage:** Depending on the stage at which the method intervenes, XAI strategies can be classified as post-hoc or ante-hoc. An ante-hoc approach involves choosing a white-box model, which inherently provides explanations by its design. An example of such models are decision trees, which offer local explanations based on the conjunction of conditions met by a specific example. Conversely, post-hoc approaches operate after a model, typically a black-box model, has made an inference. It is important to assess whether the XAI method is applicable, as some methods are specific to certain types of models while others are model-agnostic.
2. **Scope:** Depending on the scope of the explanation generated by the XAI method, methods can be classified into local or global scopes. A local XAI

method produces an explanation specific to an individual example. For instance, LIME [251], which we will discuss later, is such a method. A global XAI method generates a comprehensive explanation for the entire model, indicating which features of the model explain each class of the target variable. Global explanation methods include feature importance, among others.

3. **Result:** Based on the type of output generated by the XAI method, these are divided into surrogate models, feature relevance, or examples. A surrogate model-based XAI method produces a model that serves as the explanation. LIME is an example of such a method, providing a surrogate model that locally approximates the original model in the vicinity of the example's representation space to generate an explanation. A feature relevance XAI method focuses on identifying relevant features to produce an explanation. Lastly, an example-based XAI method constructs explanations using examples of outcomes provided by the model.
4. **Functioning:** Depending on how the XAI method operates, it can be based on perturbations or structure leveraging. A perturbation-based XAI method generates variations of an example to evaluate the effect of these variations on the outcome. Structure leveraging methods use the structure of the data or model to build an explanation. A popular way to do this is by examining gradients in a neural network, as they can provide insight into the importance of individual input values. Another approach is simplifying an architecture, such as modifying a function within the model. Examples of these XAI methods, known as architecture modification, include modifications applied to convolutional networks, like replacing max pooling with average pooling. Another category within functioning involves constructing meta-explanations, which are generated by combining explanations from other XAI methods.

The categories of this taxonomy are not mutually exclusive. For instance, LIME [251] is a post-hoc, model-agnostic method that produces local explanations based on a surrogate model operating through perturbations. Moreover, XAI methods can generate explanations in various output formats, including rules, text-based explanations, visual explanations, mixed explanations (visual + text), argument-based explanations, or even model-based explanations.

7.2 XAI methods

XAI is an active area of research with a diverse array of methods. In this context, we will explain three widely used methods that exemplify the strategies employed to enhance the algorithmic transparency of AI systems: LIME [251], Grad-CAM [269], and Shapley Values [326].

7.2.1 Grad-CAM

As we illustrated in the previous section, neural networks introduce algorithmic opacity because their design makes it difficult to trace the effect of each feature in the input vector on the output. We mentioned that the mechanism of generating outputs from inputs, which performs aggregation operations on the input vector components and uses nonlinear activation functions, makes it complex to track a feature's influence on the output. Moreover, in the case of deep neural networks, this aggregation mechanism is applied multiple times, complicating the task of producing an explanation even further. Grad-CAM introduces a technique for producing visual explanations for deep networks [269]. This mechanism is based on using the gradients of any target concept flowing to the last convolutional layer of the network to create a localization map of the input regions that predict the concept. Grad-CAM was specifically designed for convolutional networks, a type of deep neural network tailored for processing images. The idea behind Grad-CAM's visualization maps is to highlight the important regions in the input image that lead to a specific output.

Grad-CAM relies on calculating the gradient of the class score, prior to the classification layer, typically implemented as a softmax layer. The gradient, which is a multivariate version of the derivative allowing calculations over vectors, is calculated with respect to the feature map activations of a convolutional layer. These gradients are backpropagated and globally average pooled over the width and height dimensions of the feature map to obtain the neuron importance weights.

The mechanism of Grad-CAM produces local explanations. For instance, given an image for which a convolutional network detects objects, we can see which regions are significant in detecting an object. The mechanism is explained in Figure 7.1.

Suppose, for example, that the network detects that our image contains a car. Using Grad-CAM, we calculate the gradient of the score for the 'car' class in the network, which is located at the last convolutional layer. Next, we compute the activation maps for this class based on the gradient. These maps, when backpropagated to the first layer, will highlight the important regions of the image. The visual explanation, in this example, will consist of measuring the correspondence between the car class and the regions of the input image. Figure 7.2 shows an example of a visual explanation based on Grad-CAM.

The visual explanation produced by Grad-CAM is a type of correspondence explanation. According to Speith's taxonomy [278], Grad-CAM is a model-specific, post-hoc method that produces local explanations and belongs to the category of leveraging structure. It is model-specific as it is specifically

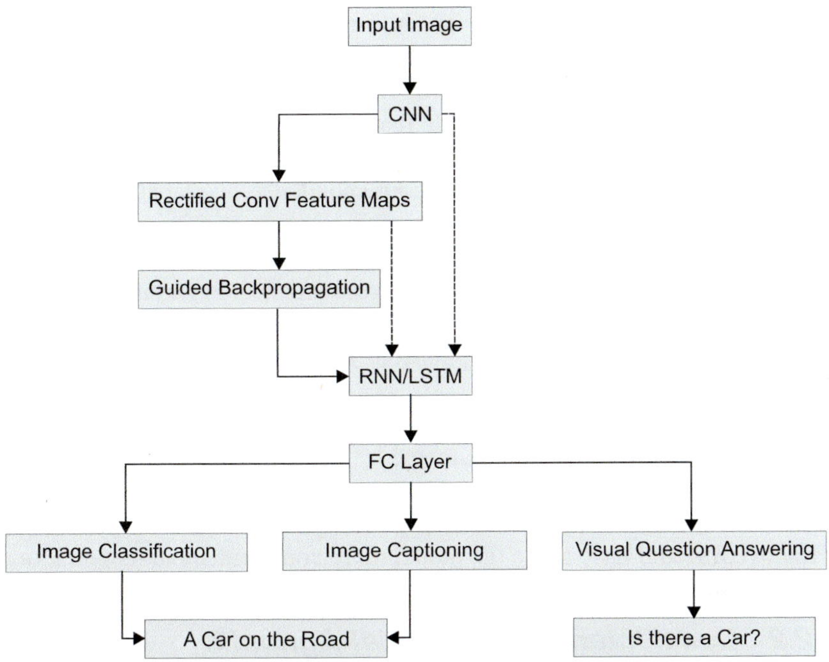

Figure 7.1: The "Grad-CAM" data flow.

Figure 7.2: The "Grad-CAM" explanation.

designed for convolutional networks. It is post-hoc since it operates during the inference phase. It is local as the explanation is generated for a single image. Moreover, it is classified under leveraging structure because it utilizes a feature of the model to highlight explainable features, in this case, the relevant regions of the input image.

Grad-CAM provides a visual correspondence explanation, meaning it establishes a correspondence between the relevant regions and the analyzed class. The method relies on the end user to rationalize the explanation by evaluating the validity of the detected correspondence. For example, we see that the highlighted region in the image contains a car, which corresponds to the class. This explanation will be plausible and thus help us evaluate why the method produced this result.

A slight modification of Grad-CAM allows for producing counterfactual explanations, i.e., assessing the outcome by removing a region from the original image. Grad-CAM accomplishes this using the negative gradient of the class score, then the network's feature maps identify regions of the input image that, if removed, would have made the model more confident about its prediction.

7.2.2 Local Interpretable Model-Agnostic Explanations (LIME)

LIME [251] is a prominent method in the field of XAI. Its goal is to identify an interpretable model based on an interpretable representation that is locally faithful, thereby approximating the behavior of the original classifier near a specific example. For text classifiers, an interpretable data representation might consist of binary vectors indicating the presence or absence of a word, even though the classifier may use more complex features, such as those encoded in deep neural networks.

LIME operates as a local explanation method. It begins by selecting an example for which it will generate an explanation. The feature vector of this example is perturbed; in the context of the text, this perturbation involves removing a word from the document. For instance, consider a text classifier trained on the 20 Newsgroups dataset, which we will refer to as model M. Suppose a news article d is classified by M under the category 'Christian'. Our goal is to understand which features of the document—in this case, words—influenced this classification. To achieve this, LIME takes the document d and generates perturbed versions by removing words. Suppose d includes the word 'god'. LIME would create a perturbed version of d, removing 'god' to get a new document, say d', and classifies it with M. If $M(d')$ = 'baseball', we use this result to indicate that d': NOT Christian. If we perturb d again, this time removing 'guns' to get d'', and $M(d'')$ = 'guns', it indicates d'': Christian. Through this perturbation mechanism, we can construct a dataset of perturbed versions of d with binary annotations indicating whether the document belongs to the 'Christian' category or not. LIME will use this dataset to train a binary classifier B. Linear classifiers are typically used for this purpose since they produce a division of the representation space into two segments, which can be used to construct an explanation. The idea is that by approximating M near d with a linear model B, we can use B to explain $M(d)$. The explanation produced using this mechanism is illustrated in Figure 7.3.

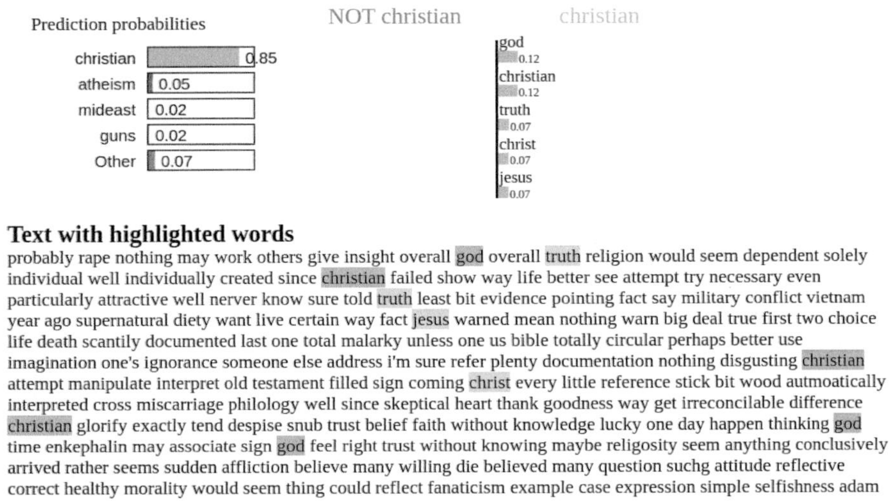

Figure 7.3: The "LIME" explanation for text classification.

The LIME graph for text at the top shows the model's confidence in classifying a text snippet as "NOT christian" or "christian". The model predicts with 85% probability that the text is "NOT christian". The two columns labelled under the categories "NOT christian" and "christian" list keywords extracted from the text along with their weights, represented by bars. These weights indicate how much each word contributes to the model's prediction. For example, the word "god" contributes 0.12 towards the prediction of "christian", which means it slightly pushes the model towards classifying the text as "christian". Below the prediction probabilities, the text is shown with certain words highlighted. These highlights correspond to the words mentioned in the word list above. This visualization helps users see which specific words in the text influenced the model's prediction.

The type of explanation obtained from B evaluates the effect of each feature. The mechanism to construct the explanation involves classifying perturbations of d using B, and measuring the distance to B. Assume that $B(d') = 0$ (meaning d' does not belong to Sports). The distance from d' to B will indicate how much a word (the word removed from d to create d') influences the outcome. The greater the distance, the larger the impact of the word on the classification.

The type of explanation produced by LIME is feature-level, indicating the relevance of each feature in the classification. Since B is binary, one way to provide a visual explanation is by using side-by-side bar plots. Additionally, words in the document can be colored according to their relevance to the classification.

LIME can also generate explanations for tabular datasets. The process is similar to that used with text: it involves perturbing an example to create a dataset of variations from the original example and then fitting a linear model around the example. This is done by using the perturbed examples and their classifications according to the original model to train the new classifier. The perturbations indicate the weight of each feature in the outcome, showing whether the classification balance leans towards one side or the other of the separating hyperplane. The impact of each perturbed example, based on its distance to the separating hyperplane, indicates the relevance of the feature in tipping the balance. Figure 7.4 presents an example of a tabular LIME explainer on a diabetes dataset, which illustrates which patient features were relevant for classifying them as healthy. The table also displays features that could potentially have tipped the balance in another direction.

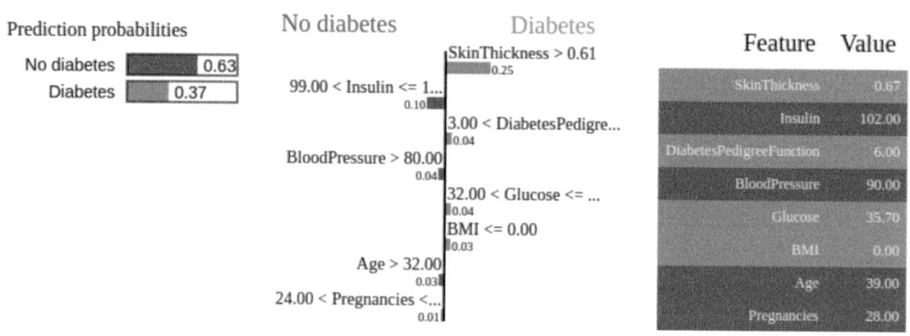

Figure 7.4: A tabular LIME explainer on a diabetes dataset illustrates which patient features were relevant for classifying them as healthy.

According to Speith's taxonomy [278], LIME is a post-hoc XAI method characterized by several key features. As a model-agnostic method, LIME can approximate any model using a linear model, meaning the approximation mechanism does not depend on the model M. Specifically, LIME constructs a surrogate model (denoted as model B) to facilitate explanation, making it an XAI method that provides outcome-based explanations.

In terms of scope, LIME is considered a local method because the surrogate model is conditioned on a specific document. Functionally, LIME is based on perturbations; it uses perturbations to generate versions of the data instance d to derive the surrogate model B. In terms of its format, LIME is recognized as a method of visual explanations that employs a surrogate model to produce explanations.

7.2.3 Shapley Additive Explanations (SHAP)

SHAP is an XAI method that assigns an importance value to each feature for a specific prediction [326]. SHAP is proposed as a framework based on a class of additive feature importance measures, which include, for example, LIME. This class of methods uses an explanatory model that is a linear function of binary variables, each indicating the presence or absence of a particular feature. SHAP shows that not only does LIME possess this characteristic, but also a traditional method of feature importance estimation known as Shapley Value Estimation. The central idea of these methods is that the effect of withholding a feature i depends on the other features in the model. Therefore, the difference in outcomes between a model that includes feature i should be measured against all possible subsets of features of the model that do not include i, denoted as $S \subset F \setminus \{i\}$. These are calculated as a weighted average of all possible differences:

$$\text{Value} = \sum_{S \subset F \setminus \{i\}} \text{Weight}(S) \times (\text{Outcome}_{\text{with } i} - \text{Outcome}_{\text{without } i})$$

This formulation allows for an indepth understanding of how each feature contributes to the prediction, considering the interactions with all other features. Given that the summation in the equation contains $2|F|$ terms, it can be approximated using sampling or by approximating the effect of removing variable i by integrating over samples from the training set. Whichever approximation strategy is employed, it eliminates the need to retrain the model and reduces the number of terms we need to calculate from $2|F|$. This approximation strategy is known as Shapley sampling values, or simply Shapley values.

Two properties are crucial in an explainable model: local accuracy and consistency. We say that an explainable model g satisfies local accuracy if g matches the output of f for the original output x. On the other hand, consistency dictates that if a model changes so that a simplified input's contribution increases or remains unchanged regardless of other inputs, that input's attribution should not decrease. Shapley values are the only set of values for additive feature importance models that satisfy both local accuracy and consistency [190].

SHAP generates explanations based on waterfall plots, which provide a visual explanation of feature importance for a given instance x. We will demonstrate how SHAP works using an example from the standard adult census income dataset from the UCI machine learning data repository (see `https://archive.ics.uci.edu/dataset/2/adult`). We will generate a waterfall plot for a single observation and calculate the Shapley values for that example. The result is shown in Figure 7.5.

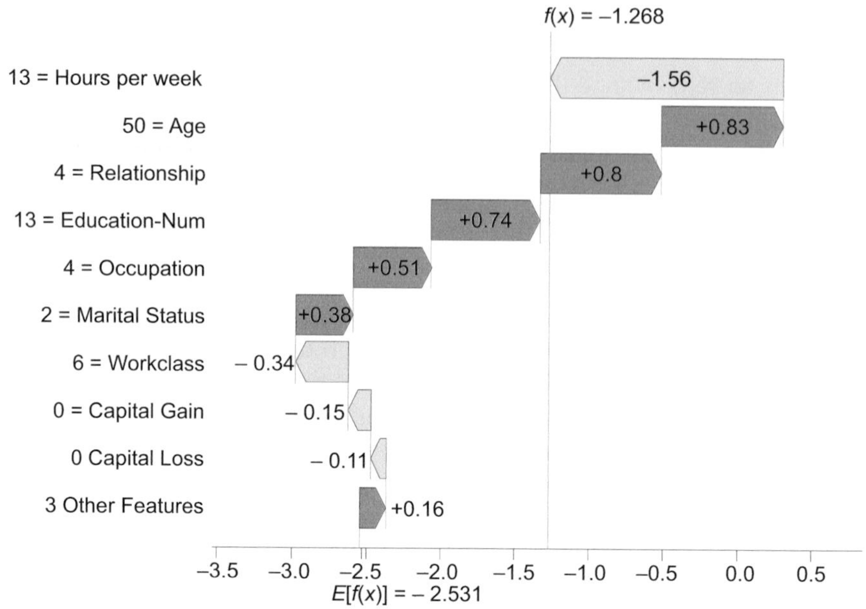

Figure 7.5: A local SHAP explainer on the standard adult census income dataset from the UCI machine learning data repository.

In the waterfall plot, the x-axis represents the values of the target variable, which in this case indicates whether the income exceeds \$50K/year. X is the observation, $f(x)$ is the value predicted by the model, and $E[f(x)]$ is the expected value of the income, i.e., the mean of all predictions generated. The Shapley value for each feature in the observation explains how much a single feature affects the prediction. This value illustrates the deviation of the predicted from the expected value and thus, explains the contribution of that feature compared to the rest of the predictions made by the model. The greater the Shapley value of a feature, the larger its contribution to the prediction. In the example, the waterfall plot displays the contributions in decreasing order. It can also be observed that the sum of all Shapley values corresponds to $E[f(x)] - f(x)$. In this example, the explanation illustrates the effects of two subsets of features. One subset, related to hours per week, workclass, capital gain, and capital loss, tends to influence the prediction in one direction. Conversely, features such as age, relationship, education, occupation, and marital status sway the decision in the opposite direction. The visualization also shows that there are three other features with a lesser impact, which are grouped into a single bin since we are using a maximum of ten features in the explanation.

SHAP also allows illustrating the effect of features on the whole dataset by overlaying the Shapley values of each example. This is done in a bee swarm plot, as shown in Figure 7.6.

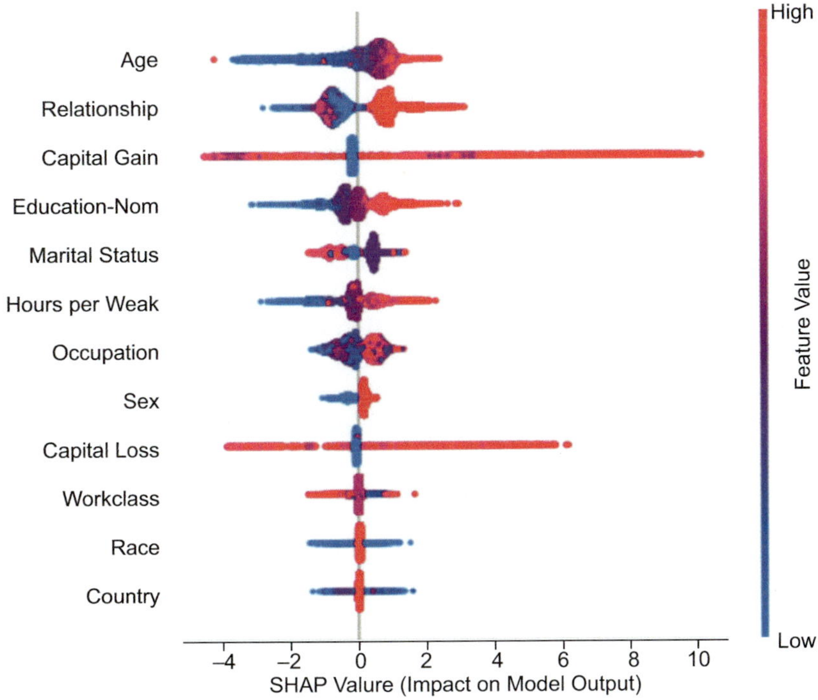

Figure 7.6: A global SHAP explainer on the standard adult census income dataset from the UCI machine learning data repository.

In this case, we observe that the explanation generated by SHAP is global, meaning it provides an explanation that does not depend on any specific example. This visualization helps to understand the impact of different features on the output of a machine learning model. Each dot in the plot represents a SHAP value for a feature for a single prediction. The position on the X-axis indicates the impact of the feature on the model's output, where values to the right suggest a higher impact and values to the left suggest a lower impact. The color of the dots indicates the feature value; red indicates high values and blue indicates low values. The features are listed on the Y-axis and sorted according to their overall impact on the model. In the plot, features such as "Age", "Relationship", and "Capital Gain" have the most significant impact on model predictions.

This type of visualization is particularly useful for interpreting complex models by showing not only which features are important, but how they affect predictions. It helps in understanding the model's behavior and can be crucial for model validation, debugging, and explaining predictions to stakeholders.

From the perspective of Speith's taxonomy [278], SHAP is classified as a post-hoc XAI method that is model-agnostic. It operates on results to construct explanations based on feature relevance. From the viewpoint of functioning, it uses the mechanism of perturbations, and in terms of scope, it can operate both locally and globally.

7.3 Further readings

Saeed and Omlin [259] provide an organized overview of the challenges in XAI, segmented by the machine learning lifecycle phases. This approach offers a roadmap for researchers to tackle existing issues and explore future opportunities. Additionally, Addadi and Berrada [2] provide an accessible entry point for both researchers and practitioners, reviewing current methodologies and underscoring the significance of transparency for AI's sustained progress. These foundational works collectively emphasize the need for XAI to foster trust and understanding in AI systems.

Several readings delve deeper into the implications of transparency and explainability in AI from various perspectives. Larsson [180] takes a socio-legal and computer science approach to advocate for a broader understanding of transparency, emphasizing its role in AI governance and regulation. Similarly, Buiten [46] focuses on the legal context, stressing the necessity of clear explanations of algorithmic biases to mitigate risks and ensure fairness in AI applications. Rogers and Howard [253] introduce a nuanced perspective by discussing the potential benefits and risks of algorithmic transparency, particularly in scenarios involving deceptive AI in human-robot interactions. These readings collectively suggest that transparency in AI is not only a technical challenge but also a critical component of ethical and regulatory considerations.

Finally, Vainio-Pekka et al. [297] systematically map the intersection of XAI and AI ethics, identifying research gaps and emphasizing the empirical and theoretical implications of XAI for ethical AI development. This text is highly recommended for those seeking to delve deeper into the relationship between transparency and the XAI approach, successfully integrating ethics into this area of study.

Chapter 8

Transformers and Generative AI

8.1 Introduction

Transformers are the cornerstone of many of the most significant current applications in AI [301]. Leveraging this architecture, AI systems have been capable of tackling complex tasks such as machine translation [244], text summarization [10], and text classification—classic tasks in NLP [77]. Beyond successfully managing these tasks, the Transformer architecture has propelled the development of Generative AI, with applications extending to chatbots, and the creation of hyper-realistic images and videos. The significant progress in Generative AI is attributed to this architecture and to additional elements that have supplemented the training of Transformer networks, such as reinforcement learning based on human feedback.

Like all major technological advances, Generative AI's progress presents enormous ethical challenges. Proper use of these technologies can bring great benefits to society, but improper use poses risks that must be analyzed. To address the challenges of modern AI, we will begin by explaining how Generative AI functions to understand its potential. Subsequently, we will explore case studies that will help illustrate the ethical challenges inherent in these technologies.

8.2 The Transformer architecture

8.2.1 Attention is all you need

The Transformer is a complex neural network architecture consisting of multiple building blocks known as Transformer blocks [301]. We will begin by describing what is contained within a Transformer block and then explore how this architecture scales by utilizing multiple base blocks.

At the core of the Transformer architecture is the self-attention mechanism. Self-attention allows the network to extract information from contexts of varying lengths. Unlike predecessor architectures, such as recurrent networks, the self-attention mechanism enables the Transformer to encode dependencies among tokens of variable length inputs, thereby operating over longer-range dependencies between input tokens.

The self-attention mechanism is implemented using feed-forward networks, and its fundamental operation is similar to that in other conventional neural networks, namely the matrix-vector product. Suppose we are processing a text input through the self-attention mechanism. The basis of this mechanism involves comparing an item of interest (a token) with other tokens in the input to reveal (and encode) its relevance in the context of the token being processed. The simplest form of comparison is the dot product.

Without loss of generality, let's assume that the self-attention mechanism follows human reading order (left to right context). This means that if we are processing the third token of the input, say x_3, we will encode this token into a new variable, say y_3. Since the Transformer in this case follows a left-to-right context, to calculate y_3 we must consider three products: x_3 with x_1, x_3 with x_2, and x_3 with x_3. These products are normalized using a softmax layer creating a weight vector α_{ij}, which indicates the relevance of input token j to the output token i. This process is expressed in the following equation:

$$y_i = \text{softmax}\left(\sum_{j=1}^{N} \alpha_{ij} x_j\right),$$

where α_{ij} is computed based on the similarity between tokens x_i and x_j.

This basic self-attention mechanism indicates that the attention weights α_{ij} are learned. They are learned because the variables x_1, x_2, and x_3 are learned, as the network does not process the tokens directly but their representations (encodings). Since x_1, x_2, and x_3 are parameters, the coefficient α_{ij} is also learned. We say that α_{ij} is the attention coefficient from i to the token j.

The Transformer enhances this basic attention mechanism by encoding each token according to three roles: query, key, and value. The query role implies that the input encoding is considered the current focus of attention when compared with preceding inputs. The key role uses a token as a preceding input, and thus, the input embedding from the previous step is compared with the next one. The value role uses the input embedding to calculate the output value at the current time step. Essentially, to compute these role vectors, the self-attention mechanism uses three parameter matrices, which calculate q_i, k_i, and v_i from x_i based on the matrix-vector product, as shown in the following matrix equations:

$$q_i = W^Q x_i, \quad k_i = W^K x_i, \quad v_i = W^V x_i.$$

Here, W^Q, W^K, and W^V are the parameter matrices for the query, key, and value roles, respectively.

Now that we have the key components of the mechanism, we can understand how y_i is computed, i.e., the encoding of x_i calculated using the self-attention mechanism. Essentially, instead of operating directly on x_i, the self-attention mechanism operates on the query, key, and value vectors as follows:

$$y_i = \sum_{j \leq i} \alpha_{ij} v_j,$$

$$\alpha_{ij} = q_i \cdot k_j.$$

These equations represent how the outputs are computed by weighing the value vectors (v) with the attention weights (α), which are determined based on the similarity between the query and the keys.

As demonstrated by these equations, the method for generating the encoding y_i involves performing query/key comparisons (products) which are used to calculate attention coefficients. These coefficients are then compared with the value vectors. The sum of these products yields the output vector y_i. The mechanism is explained in the diagram shown in Figure 8.1.

An important aspect of this mechanism is that the output vector is computed based on a linear combination of the value vectors from the input, with attention coefficients calculated by the key-query comparison mechanism. To ensure that the combination of these factors is indeed linear, the attention coefficients are normalized using a softmax layer:

$$\text{Softmax-layer}(Q, K) = \text{logistic}\left(\frac{QK^T}{\sqrt{d_k}}\right),$$

where $\text{logistic}(z_i) = \frac{e^{z_i}}{\sum_j z_j}$ and d_k is the dimensionality of the vectors q and k. This normalization step ensures that the coefficients sum to one, allowing for a proper linear combination of the value vectors to form the output vector.

A key feature of the Transformer is that it does not compute the encodings sequentially but can do so in parallel. This enables efficient processing of the inputs. However, it imposes a limitation that the input length must be fixed, determined by the number of input tokens that can be processed simultaneously through the self-attention mechanism. This is an architectural hyperparameter of the network.

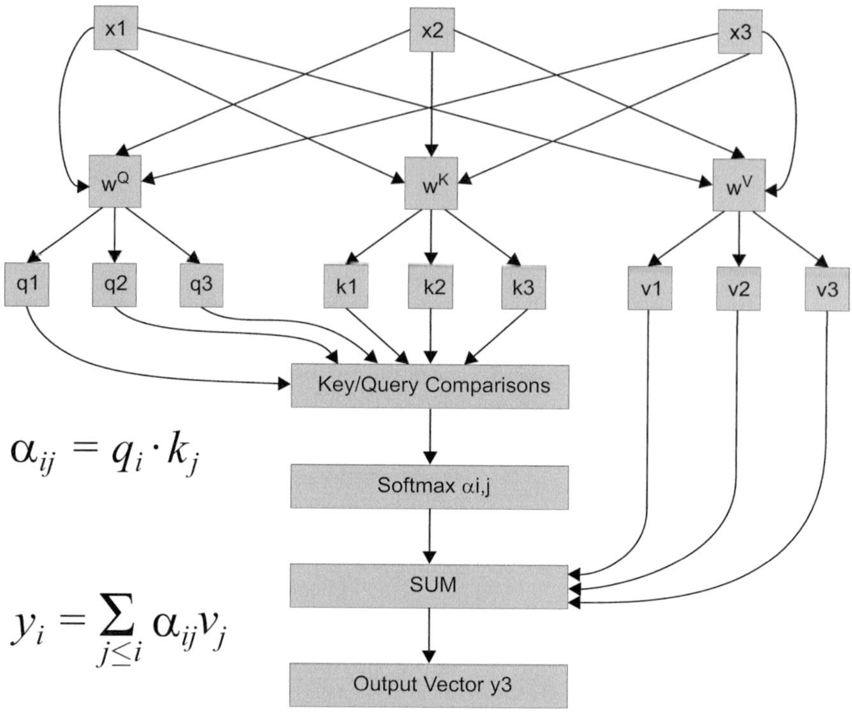

Figure 8.1: The self-attention mechanism of the transformer.

The relevance of the self-attention mechanism is substantial. This mechanism allows the Transformer to encode long-range dependencies between input symbols. Specifically, depending on how the Transformer is trained, the network will learn embeddings that are tailor-made for the required tasks, with a significant capacity to encode and utilize dependencies of arbitrary length on the input.

A transformer block is based on the self-attention mechanism. However, it includes other fundamental operations to encode input symbols. After passing through the self-attention layer, a residual connection is applied that combines the output vector with the input. This residual connection is based on the sum of both vectors, which are then normalized to ensure that the sums operate within a bounded domain. Finally, the vectors from the Add and Normalize layer feed into a feed-forward layer. Both the input encoding of the feed-forward network and its output are combined again in a second residual connection, using addition followed by normalization. Figure 8.2 illustrates the sequence of operations that make up the transformer block.

We are nearly finished describing the components within a transformer block. An additional element incorporated into the architecture is multi-head attention. Instead of operating on a single self-attention layer, transformers utilize multiple such layers in parallel, each with its own parameters for calculating the query,

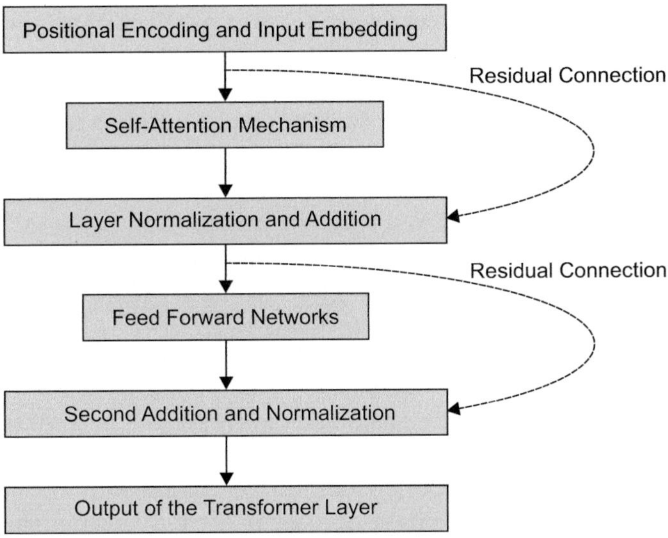

Figure 8.2: The Transformer block begins with positional encodings and input embeddings, which are then passed to the self-attention mechanism.

key, and value vectors. The outputs from these parallel layers are concatenated into a single vector, which is then fed into a feed-forward layer that reduces the vector's dimensionality. The purpose of the multi-head attention mechanism is to scale the transformer's capabilities based on the self-attention mechanism, by adding more learnable parameters and thus enhancing the transformer's encoding capacity.

Another key component of the Transformer architecture is positional encodings. The parameter matrices used to compute the query, key, and value vectors are shared across different input tokens. Consequently, transformers have the property of being permutation-equivariant with respect to the order of input tokens, meaning that swapping the order of tokens does not alter the outcome. To encode the order of tokens, a positional encoding vector is constructed for each input position. These encodings are combined with the token encodings, helping to preserve both the token encoding and its position simultaneously. Various techniques are used to construct positional encodings, with those based on sinusoidal functions being the most commonly employed.

8.2.2 Transformer encoder-decoder

Transformer blocks can be stacked one after another to create increasingly higher-level encodings. If the input to the first transformer block is a sequence of data tokens, we say that the transformer operates as a transformer encoder. It is possible to place a linear layer followed by a softmax layer over the symbol vocabulary at the output of the last block in the transformer chain. In the case of text, this last softmax layer sweeps the token vocabulary. What the transformer delivers in this last layer are the output probabilities of the network over the tokens. We refer to this symbol production strategy as a transformer decoder.

To aid symbol generation, the decoder operates using a token feed strategy called auto-regressive generation, in which the input token at position i produces the output token at position $i+1$. By shifting the token generation by one position, the task addressed by the transformer is called **next token prediction**. This is the basis for text generation in transformer-type networks.

Both transformer operation strategies, encoding and decoding, can be used simultaneously. The key idea used to connect both modes of operation involves adding a layer called encoder-decoder attention, which uses the encoding from the transformer encoder and combines it with the encoding from the decoder. While the decoder is operating autoregressively for the next token prediction, each of the symbols from the generator is also connected with the encodings from the encoder. This type of conditional generation with attention is called cross-attention, as the transformer decoder is not only paying attention to the input tokens but also to the encodings coming from the transformer encoder. This strategy allows correlating two sequences of tokens, the input sequence of the encoder and the input sequence of the decoder, a mode of operation on the architecture also known as transformer seq2seq. While the first operates in encoder mode to condition the decoder, the transformer decoder operates in auto-regressive mode to generate an output token sequence from the input token sequence of the decoder. This feeding method is the basis of machine translation algorithms, where we use a token sequence in the encoder in a source language and place a second token sequence in the decoder, in auto-regressive mode, in the target language. If we do this with many pairs of sequences, which involves having a dataset of aligned sentences, the transformer will learn to generate the token sequence in the target language from the sequence in the source language. In inference, when we only have the text in the source language, the decoder will operate based on auto-regressive generation, starting from the first symbol of the sequence, which by default is always the START symbol.

8.2.3 Transformer encoder

The Transformer architecture has proven extremely useful for tackling various tasks in NLP. One notable implementation involves using only the Transformer encoder for the masked language model task. This involves connecting a linear (feed-forward) layer to the output of the last Transformer block, followed by a softmax layer to span the vocabulary of tokens. The masked language model task predicts an output token from an input sequence where some tokens have been masked. This forms the foundation of Bidirectional Encoder Representations from Transformers (BERT) [77], which builds word encodings by training the Transformer on a large text volume using the masked language model task.

BERT was trained on a diverse, extensive text corpus, including BookCorpus, which contains over 800 million words, and Wikipedia, totaling 2.5 billion words. Utilizing large text volumes provides these models with a greater generalization capability. However, it is important to note that the model will replicate biases present in these texts, which were not preprocessed to address biases. The model processes the texts as they are.

Once a Transformer encoder is trained using the BERT strategy, we can input a new sentence during inference. By performing a forward pass over the network, we can retrieve the word encodings from the Transformer blocks, using these vectors as context-dependent word embeddings.

Word embeddings are extremely useful as they help us build vector representations of documents from the word embeddings by combining the vectors using an aggregation operation such as averaging, a strategy known as average word embedding. Based on these vectors, we can train document classifiers. There are also other ways to utilize BERT vectors, employing sentence-level aggregation strategies known as Sentence BERT, which provide sentence embeddings for sentence-level tasks.

BERT has also been used as a pretrained model. Once the Transformer encoder is trained using BERT, we can fine-tune this network for downstream tasks by feeding it input and output from a specifically annotated dataset. This approach, known as fine-tuning, has successfully evaluated BERT in various NLP tasks, including document classification, sentiment analysis, and closed question and answering.

While the use of word embeddings offers significant benefits, it also presents some risks. Pretrained models like BERT rely on large text volumes that contain biases. We will see that analyzes based on word embeddings illustrate the effects these biases have on the learned representations. Our strategies will demonstrate that the analogies drawn from word embeddings reveal the reproduction of cultural and historical biases.

8.2.4 Transformer decoder

Just as the transformer encoder has been used to construct representations, the transformer decoder has been utilized for generation. The transformer decoder forms the basis of what we know as Generative AI. Relying on the auto-regressive generation mechanism, the transformer decoder can be used to train generative models. The latest advancements in generative AI are largely due to the adoption of this architecture.

The transformer decoder can operate without the encoder in an auto-regressive text generation mode. Based on the task of next token prediction, the transformer decoder employs a variant of the masked language model called the causal language model. The strategy here involves modifying the self-attention layer to only focus on preceding tokens in the input sequence. Unlike the masked language model, which considers context both to the left and right of a symbol—hence this is why BERT is a bidirectional transformer—a causal language model only pays attention to preceding tokens. By forcing the attention mechanism to act causally, the decoder is prepared to generate text using this mechanism combined with the auto-regressive generation strategy.

A transformer decoder may use several transformer blocks. Similar to the encoder, the transformer decoder uses positional embeddings at the input and softmax at the output. This final layer produces output probabilities over the token vocabulary. When trained on the causal language model task, the decoder learns to produce texts based on next token prediction. This text completion task is the foundation of LLMs.

A transformer decoder trained for text completion on a large volume of text can generate text based on input text. Like BERT, the early LLMs based on the transformer decoder were trained using the BooksCorpus dataset. This led to the development of GPT-1, the first model from OpenAI known by its acronym Generative Pretrained Transformer [242]. The original paper on GPT-1 employs fine-tuning on the base model to assess GPT-1's capabilities in downstream tasks. Like with BERT, fine-tuning a pretrained transformer significantly improved various NLP tasks, including closed question and answering, semantic similarity, and text classification.

8.3 Collecting human feedback for the transformer

The transformer architecture represents a disruptive innovation in AI. While its self-attention mechanism and the opportunity it provides for training models using unsupervised strategies are notable features, its connection with reinforcement learning has likely garnered the most attention in recent years. In specific, the relationship between the Transformer architecture and Reinforcement Learning based on Human Feedback (RLHF) represents a significant development in the field of AI, particularly in NLP.

The collection of human feedback has proven to be extremely valuable in working with LLMs, particularly for tasks involving text such as summary generation. Stiennon et al. [281] incorporated human feedback on Reddit posts by applying various automatic strategies to generate summaries for a given post. Once a list of candidate summaries was available, two were selected for human evaluation. The human annotator then judged which summary better represented the original post, allowing the creation of a reward model. For this process, two summaries of one post are input into the reward model, which then calculates a reward for each summary. The model's loss is determined based on these rewards and human labels, which are then used to update the reward model.

We will utilize the reward model to supervise the fine-tuning of a transformer on the task of text summarization. The process involves sampling a new post from the dataset and providing a summary generated by our model. The reward model then assigns a reward to this summary, which is used to further refine our model.

The principle of using the reward model for model alignment based on human feedback led to the development of Instruct GPT [234]. Instruct GPT is an LLM designed to generate outputs conditioned on prompts, which can include instructions, contextual information, or examples of the task to be solved. It uses GPT-3 as its foundational model, an enhanced version of GPT-2 with significantly more parameters and trained on a larger text corpus.

Instruct GPT was aligned using a dataset of prompts encompassing various tasks. When a prompt from this dataset is sampled, a human labeler demonstrates the desired output behavior by writing the expected response. This data is then used to align GPT-3. In a subsequent phase, prompts and several model outputs are sampled and a human labeler ranks these outputs from best to worst. This data trains a reward model. Once the reward model is trained, it optimizes the model based on the calculated rewards. A new prompt is sampled, the model generates an output, and the reward model evaluates this output. The resulting reward is used to update the model.

Instruct GPT includes improvements such as a broader range of tasks in its prompts dataset, not limited to summaries. These tasks include open-ended text generation, open question and answering, brainstorming, chat, rephrasing, summarization, classification, closed question and answering (multiple choice), and text extraction. While the foundational model achieves interesting results on these tasks, demonstrating the model's text processing capabilities, the model aligned with the prompts dataset shows significantly improved results. This enhancement relies heavily on aligning model outputs with prompts.

Instruct GPT forms the basis of ChatGPT, an impressive OpenAI chatbot capable of processing prompts and producing aligned responses. ChatGPT is built on GPT3.5, a foundational model that extends the capabilities of GPT-3. However, OpenAI has updated its foundational models, releasing GPT-4 in 2023, their most advanced LLM to date. The capabilities of ChatGPT based on GPT-4 have been so remarkable that it has been described as moving towards a general AI, i.e., an AI capable of tackling new tasks for which it was not specifically trained. The ethical implications and risks of this technology will be the main focus of our next chapter.

8.4 Further readings

The Transformer architecture, particularly in the form of ChatGPT, has profoundly influenced various sectors, including education, healthcare, and customer service. ChatGPT has garnered significant attention in education due to its potential to reshape educational norms and offer personalized, adaptive learning experiences. However, there are concerns regarding its potential to diminish analytical skills and encourage misconduct [117, 261, 307].

Within healthcare, the latest iteration, GPT-4, is noted for its expanded knowledge base and enhanced problem-solving capabilities. Notably, its new ability to analyze images holds promise for medical image interpretation and diagnosis [305]. Moreover, ChatGPT has been assessed for its ability to respond to complex clinical queries, showing potential as an interactive tool for medical education [187]. In customer service, ChatGPT has been employed to provide logic and informational context across the majority of responses, demonstrating its applicability in this area [54].

Nonetheless, there are ethical concerns associated with the use of ChatGPT, particularly regarding the potential for spreading misinformation, issues surrounding privacy, and the risk of overreliance on technology [123]. Future research is necessary to address these limitations and potential risks, with an emphasis on the careful consideration and verification of the information provided [54].

In summary, while the Transformer architecture, as exemplified by ChatGPT, has the potential to revolutionize various industries, it is essential to consider its limitations and ethical implications carefully to ensure its responsible integration into these domains.

Chapter 9

NLP and Representational Bias

9.1 Introduction

In NLP, bias has been a persistent challenge, with its roots often traced to the representations produced by word embeddings. These word embeddings, which serve as vector-based representations of words in semantic space, are susceptible to bias because they reflect the textual data from which they are derived. Bias, in this context, is typically defined by the negative associations that emerge within these word representations, influencing how words are positioned relative to one another. Through the use of analogies, such as the well-known "Man is to computer as woman is to homemaker," [36] researchers have illustrated the extent of bias present in word embeddings and evaluated its potential downstream impact on various NLP tasks using analogies, the presence of bias in these representations is illustrated, and its potential impact on downstream tasks is assessed [320].

As discussed in the previous chapter, the Transformer architecture is currently dominant. However, prior to the Transformer, the prevalent techniques in NLP for representation learning primarily relied on context-independent word embeddings. These vectors were computed either through feed-forward networks (word2vec) or via matrix factorization strategies of the term-document matrix from a text corpus (GloVe). These vectors represented the position of a word in the semantic representation space. Proximity between two words indicated a semantic connection, while distance suggested a weak semantic link. In both strategies, vectors were retrieved from a model pre-trained on a large volume of text. The use of corpora to train these models is based on the principle of textual proximity: if two words tend to co-occur in the text, then these words are connected. By using text to establish semantic connections between words, word embeddings reproduce biases and stereotypes because the text describes the world as it is, thus reproducing existing biases.

The analysis of bias in NLP has been approached from various perspectives, for example, strategies for quantifying biases based on metrics have been established. Likewise, bias mitigation strategies have been proposed, most of which are post-hoc strategies that make adjustments to word embeddings to achieve a specific objective. Most of these studies focus on analysis based on disadvantaged groups, with a strong emphasis on gender analysis. In terms of attributes of interest, the reproduction of stereotypes in professions stands out. There is also a growing interest in studying the effect of counterfactuals in text, proposed as a strategy for mitigation and corpus analysis that would allow working on texts with fewer biases and stereotypes. We will begin by illustrating the issues in this area based on a pioneering work in the analysis of biases in NLP: Man is to computer as woman is to homemaker.

9.2 Word analogies and stereotypes

9.2.1 Hard debias

Bolukbasi et al. [35] demonstrate that word embeddings trained on corpora used to construct word representations, such as GloVe, exhibit gender stereotypes for male and female. They introduce a key concept based on a geometric idea, aiming for an ideal representation from which to aspire to a debiased model. This concept hinges on distinguishing between gender-neutral words and gender-biased words, noting that gender-neutral words are linearly separable from gender definition words in the word embedding space. Based on these properties, they propose a methodology to modify an embedding by removing gender stereotypes, eliminating stereotypical associations while preserving descriptive associations between words.

The analysis begins by defining what a word analogy is. Given three words, for example, "he," "she," and "king," we look for a fourth word to complete the analogy: "he is to king as she is to x." Word embeddings such as GloVe or Word2Vec find astonishing analogies. While for this example, most pretrained models will solve x as "queen," in other analogies involving professions, the outcomes differ significantly. For instance, the analogy "man is to doctor as woman is to x" is generally resolved as "nurse." Similarly, "man is to computer programmer as woman is to x" typically results in "homemaker." These analogies clearly reproduce gender stereotypes.

Bolukbasi et al. [35] introduce a method called hard-debias, which is based on identifying the gender subspace. The method starts with a set of words to neutralize and works with the word embeddings from a word set, which are descriptive of gender. For each word set, the mean of the word embeddings that comprise it is defined, denoted as u_i. Then, the bias subspace B is defined as the first k rows of the Singular Value Decomposition (SVD) of the matrix of deviation vectors around the mean for all the word embeddings that make up the word sets. If the word sets describe gender words, for instance, having two-word sets—one for male and one for female—B will allow the identification of the gender subspace of the corpus we are working with.

After determining the gender subspace, work is done on words to be neutralized, such as professional words. For each word to be neutralized, a new embedding, which the authors call re-embedding, denoted as w, corrects the original word embedding by the mean vector of the gender subspace, which we denote as $\vec{w}_B$. Then, for each word in the set of words to be neutralized, we calculate the mean over the group based on the re-embedded vectors, denoted as μ. Next, the vector that measures the difference between the group's mean and the mean vector of the gender subspace is calculated, i.e., $v = \mu - \mu_B$, where μ_B represents the mean vector of B. Finally, for each word to be neutralized, the word embedding is calculated as:

$$\vec{w} = v + \sqrt{1 - \| v \|^2} \frac{\vec{w}_B - \mu_B}{\| \vec{w}_B - \mu_B \|}.$$

The effect of the Hard Debias method is based on zeroing out the gender projection of each word along a predefined gender direction. This method of recomputing embeddings assumes that a fair condition for representation is based on the symmetry of gender-sensitive words, such as professions, around the gender axis. It is a simple yet key geometric idea. The representation space is fair if the words to be neutralized, such as professions, are gender-neutral. This means that, within the gender subspace, the words to be neutralized should be very close to each other, to prevent the reflection of gender stereotypes in professions.

The concept derived from Bolukbasi's idea entails an implicit definition of what is considered gender neutral [35]. According to the Hard debias approach, there is no gender bias if each word in the vocabulary that is not explicitly gendered is equidistant from both elements of all explicitly gendered word pairs. Consequently, **fairness in the representation space is akin to the concept of symmetry**.

9.2.2 Lipstick on a pig: Debiasing methods do not remove them

Gonen & Goldberg [336] analyzed the Hard debias method and subsequent approaches, all based on the concept of symmetry and neutrality concerning the gender direction in the representation space. Through clustering experiments in the word embeddings space, they revealed that Hard debias was unable to neutralize associations between words due to residual implicit bias in the components of the representation not aligned with the gender subspace. This issue arises because the proximity relationships encoding gender bias are not confined to the gender subspace but persist beyond it. This persistence is due to the projection bias on which Hard debias relies, correlating with bias by neighbors, thus maintaining the proximity relationships of the original space based on neighborhood structure. Clustering of gender words on Hard debiased word embeddings of gendered words does not effectively reveal the bias. For example, "nurse" is no longer close to explicitly marked feminine words. However, the bias still manifests with the word being close to socially-marked feminine words, such as "hairdresser" or "captain." This suggests that an effective way to measure bias is not just based on symmetry concerning the gender direction but also on the percentage of male/female socially-biased words among the k nearest neighbors of the target word.

The metaphor "lipstick on a pig," illustrating that debiased methods based on symmetry only superficially remove bias, shows that words with strong gender bias are easy to cluster together. As such, associations between stereotyped words can persist in debias methods based on symmetry because they preserve structural associations of the representation space based on neighborhoods. It also illustrates that words with implicit gender from social stereotypes (e.g., "hairdresser" or "captain") still tend to group with other implicit-gender words of the same gender, similar to non-debiased word embeddings. Thus, the implicit gender of words with prevalent previous bias is easily predictable based on their vectors alone.

9.2.3 Limitations of gender neutrality subspace-based methods

Readers may notice several limitations of debiasing methods based on the concept of symmetry. Firstly, many of these methods, notably the pioneering Hard DeBias, rely on the definition of word sets. The selection of word sets and words to be neutralized is a delicate process, as bias can be inadvertently introduced during this selection. Essentially, this raises the question of who decides which words should be neutralized and which words represent a societal axis that needs protection. Such definitions inherently carry all the risks associated with human-made definitions, reproducing biases and stereotypes through selection bias.

A second limitation is that these methods are based on the premise that word embeddings are context-independent. This implies that a word's representation is static, and therefore, it maintains the same representation regardless of the context in which it is used. Methods like Word2vec and GloVe operate on this principle. The issue with context-independent word embeddings is that they do not account for polysemy—a word can have multiple meanings, and the appropriate meaning is contingent upon the sentence in which the word is used. Thus, by ignoring the context dependency, we hinder the ability to capture an appropriate representation of the word in the context it is used, thereby neglecting the effects of polysemy. It is preferable, then, for a word's representation to be context-dependent to address polysemy effectively. This is what BERT achieves, as it allows for the retrieval of vectors conditioned on the sentence in which the word is used.

9.3 Counterfactual data augmentation

While methods based on symmetry and correction of word embeddings are inherently post-hoc—meaning they operate on precomputed representations and then intervene to make corrections—it is also possible to work on the data before it is used to train a model. This approach is based on the assumption that data, as it is, reproduces biases and stereotypes. In this sense, raw data represents the world as it is, complete with all its asymmetries and biases. Therefore, an appropriate way to reduce biases involves modifying the data so that it represents the world as it should be.

Counterfactual Data Augmentation (CDA) aims to create alternative versions of data that reduce the inherent biases in the data. The seminal idea of these strategies is that by detecting a stereotypical mention in a sentence, we can mitigate its potentially harmful effect by providing the dataset with alternative versions that balance the mention across different groups. For example, consider the source sentence "A man is walking." Since "man" is a word that belongs to a set of male-oriented words, we can balance the effect of this sentence to prevent the action of walking from being solely associated with men. By replacing "man" with "woman," we generate a new sentence, "A woman is walking," which balances the gender effect in the data.

One advantage of these data augmentation techniques is that they are model-agnostic, meaning they can operate over any text encoding model, including models that consider the context of a word like the transformer. This was precisely what Webster et al. [311] did, using various data augmentation strategies to modify the training datasets for BERT. Initially, the authors demonstrate that both BERT and its distilled version ALBERT learn and utilize gendered correlations. Accordingly, they applied counterfactual data augmentation to gender mentions to reduce gender correlations. To apply counterfactual pretraining to BERT, the authors generate supplemental training examples from English Wikipedia using gendered word pairs (e.g., he – she). First, they identify sentences containing one of the gendered words and then generate the counterfactual sentence by substituting the gender-partner of the word in its place. Results on both BERT and ALBERT show that gender-based correlations decrease using this technique. Another important finding of the study highlights that these models are resilient to fine-tuning; that is, after reducing gender-based correlations and using the pretrained model for a new task, the gender correlations remain low. An interesting aspect of the strategy is that it maintains the accuracy of the pretrained model when used in downstream tasks. This reveals that we can reduce bias in pretrained models without assuming costs in terms of the utility of the model.

9.4 Limitations of model debias strategies

While methods based on symmetry and correction of word embeddings are inherently post-hoc, meaning they operate on precomputed representations and then intervene to make corrections, it is also possible to work on the data before it is used to train a model. This approach is based on the assumption that the data, as it is, reproduces biases and stereotypes. Thus, raw data represents the world as it is, with all its asymmetries and biases. Therefore, a suitable way to reduce biases is to modify the data so that it represents the world as it should be.

Counterfactual Data Augmentation (CDA) aims to create alternative versions of the data that reduce the inherent biases. The seminal idea of these strategies is that by detecting a stereotypical mention in a sentence, we can counteract its potential harmful effect by providing the dataset with alternative versions that balance the mention across other groups. For example, consider the source sentence "A man is walking". Since "man" belongs to a word set composed of male terms, we can balance the effect of this sentence to prevent the action of walking from being solely associated with men. By replacing "man" with "woman", we obtain a new sentence, "A woman is walking", which balances the effect in the data of a mention oriented only to one gender group.

An advantage of these data augmentation techniques is that they are model-agnostic, and thus, can operate on any text encoding model, including context-aware models like the transformer. This is precisely what Webster et al. [311] did, who used different data augmentation strategies to modify the training datasets for BERT. First, the authors show that both BERT and its distilled version ALBERT learn and utilize gendered correlations. Accordingly, they use counterfactual data augmentation on gender mentions to reduce gender correlations. To apply counterfactual pretraining to BERT, the authors generate supplemental training examples from English Wikipedia using gendered word pairs (e.g., he – she). They first find sentences containing one of the gendered words and then generate the counterfactual sentence by substituting the word's gender partner in its place. Results on both BERT and ALBERT show that gender-based correlations decrease using this technique. Another significant finding of the study highlights that these models are resilient to fine-tuning, meaning that after reducing gender-based correlations and using the pretrained model for a new task, the gender-based correlations remain low. Another interesting aspect of the strategy is that it maintains the accuracy of the pretrained model when used in downstream tasks. This underscores the possibility of reducing bias in pretrained models without incurring costs in terms of model utility.

9.5 Further readings

Debiasing methods in NLP have traditionally focused on isolating or removing information related to sensitive attributes. However, there is an increasing argument for the 'fair' utilization of this sensitive information, supported by explanations rather than its blanket removal [192]. This perspective suggests that sensitive attributes should be used thoughtfully, with an emphasis on transparency and fairness in the processing of data. In addition, the integration of interactive setups that incorporate user feedback has been proposed as a means to achieve a more balanced and fair approach to bias mitigation. This method not only improves task performance but also enhances the reduction of bias in the explanations provided by models, all while maintaining the accuracy of predictions [233].

Furthermore, strategies that are independent of specific models, known as model-agnostic debiasing strategies, have been developed to strengthen NLP models against a variety of adversarial attacks. These strategies are designed to either maintain or enhance the generalization capabilities of the models, ensuring that they remain robust across different tasks and datasets [153]. A two-stage pipeline has also been introduced to address biases in pre-trained language models, focusing on reducing biases in both internal and downstream contexts, while preserving the models' expressive power [184]. This is complemented by a novel framework that examines bias in pre-trained transformer-based language models through movement pruning, offering new insights into gender bias and proposing improvements to existing debiasing methods [317].

In addition to these approaches, it has been recognized that many debiasing methods neglect the interaction between multiple societal biases. To address this, a new debiasing model has been proposed, which utilizes the synergy between various societal biases to simultaneously mitigate multiple biases [161]. Moreover, a method has been introduced for debiasing contrastive learning, aiming to alleviate biased latent features and reduce their presence in the model's representations [192]. Lastly, the relationship between extrinsic and intrinsic bias in NLP models remains a relatively unexplored area. A new framework has been proposed to measure both types of bias simultaneously, offering a more comprehensive perspective on bias in NLP models [191].

ADVANCED TOPICS IV

Chapter 10

Benefits and Risks of LLMs

10.1 Introduction

The remarkable advancements in LLMs such as GPT-4 have transformed the landscape of AI, showcasing unprecedented capabilities in understanding and generating human-like text. From their early iterations, like GPT-1, to more sophisticated models such as ChatGPT, these systems have demonstrated an ability to tackle an impressive array of tasks, from writing essays to generating code, often with minimal instruction. Yet, these breakthroughs in AI technology raise profound ethical concerns. While LLMs offer transformative potential in fields like education, healthcare, and customer service, their widespread adoption also brings questions about their reliability, transparency, and societal impact.

A key capability of LLMs is their proficiency in few-shot learning, where the models can perform tasks they have never been explicitly trained for with competitive accuracy. This adaptability represents a step toward general AI, a long-standing goal in the AI community. However, the ability of LLMs to generalize across diverse domains without fine-tuning also introduces ethical concerns regarding misuse and unintended consequences. Without rigorous guardrails, LLMs could be deployed in critical areas—such as legal advice or mental health support—where incorrect or biased outputs could cause significant harm. This flexibility, while impressive, demands an ethical framework that addresses the potential risks of using such technology in high-stakes environments.

Moreover, the sheer scale of these models, trained on trillions of parameters and vast amounts of text, introduces additional risks related to transparency, accountability, and data privacy. As LLMs are integrated into more societal systems, the opacity of how they make decisions—coupled with their reliance on potentially biased or outdated training data—presents significant ethical dilemmas. The rapid pace of AI progress outpaces the development of regulatory frameworks, raising concerns about the responsible governance of these technologies.

One of the most impressive capabilities of LLMs is their ability to solve new tasks for which they have not been trained without needing to adjust the model's parameters to the new task. This capability is known as few-shot learning. Advanced LLMs are few-shot learners [44], meaning they can tackle new tasks with competitive performance compared to a model fine-tuned for that specific task. This adaptability of LLMs to new tasks suggests that this type of AI is the first to achieve a significant breakthrough in terms of approaching general AI [45]. While much of this advancement is due to these models' enormous capacities, with trillions of parameters, trained on vast volumes of text, and aligned to process instructions based on massive human feedback, it is still impressive that these models continue to improve rapidly. However, despite these vast advancements, there are inherent risks associated with the rapid progress of these language technologies. We will address these in the following sections, leading to a reflection on the ethical aspects, uses, and abuses of LLMs.

10.2 Hallucinations in LLMs

LLMs often exhibit a tendency to produce hallucinations, resulting in outputs that are inconsistent with real-world facts [155]. This phenomenon poses significant challenges since reliable results are essential for their application in various tasks. As highlighted by Huang et al. [143] in their recent survey on the subject, we have identified different types of hallucinations in LLMs.

The first type we discuss is "factuality hallucination," where LLMs occasionally generate outputs that are inconsistent with real-world facts or potentially misleading. This can be due to factual inconsistencies, where the model provides a factually grounded but inaccurate or contradictory output. This includes errors such as attributing events to incorrect figures, date inaccuracies, and other historical errors. Factuality hallucinations can also arise from factual fabrication, where the model creates data about non-verifiable information, including urban myths or conspiracy theories. Both factual inconsistencies and fabrications highlight the risks of using LLMs as a data source.

Another type of hallucination, known as "faithfulness hallucinations," involves errors in the logic of aligning question and answer. These inconsistencies show that LLMs sometimes fail to process the prompt properly, resulting in a response poorly aligned with user instructions. We recognize three types of faithfulness hallucinations:

1. Instruction inconsistency, where the output is poorly related to the user's intent, including failures to follow or misunderstand instructions.
2. Context inconsistencies, where a user provides a real-world fact in the prompt's context, and the output contradicts this statement.
3. Logical inconsistency, involving errors in mathematical operations or the application of logical principles, indicative of a contradiction in a reasoning task.

The causes of these hallucinations are varied, with one significant factor being the use of inconsistent data sources during model pre-training. Factors that can exacerbate hallucinations include inadvertently included misinformation in the training data. Since LLMs generate outputs based on the data they have processed during training, these causes relate to imitative falsehoods—essentially, if the data contains falsehoods, the LLM will reproduce them. Another source of hallucinations can be the introduction of social and historical biases from data sources. This includes "duplication bias," where repeated facts in the data can lead LLMs to shift from generalization to memorization, prioritizing the recall of this data. Additionally, the presence of social and historical biases in the data can be perpetuated through stereotypical associations.

One cause of hallucinations in LLMs is due to the knowledge boundary. LLMs have limited capabilities to handle up-to-date information. Outdated factual knowledge presents challenges, as foundational models have a temporal boundary and can become outdated over time. Prompts that exceed these boundaries may force the models to provide answers that could result from fabricating facts beyond the model's temporal edges. Another limitation of these models relates to domain knowledge deficiency, which is the lack of handling concepts associated with specific domains. This issue arises because LLMs are predominantly trained on datasets of general knowledge, and therefore, their ability to respond to questions related to specific domains encounters the model's domain boundary.

Further causes of hallucinations include 'inferior data utilization,' where spurious correlations captured during training lead to factual inaccuracies. This is often a result of 'knowledge shortcuts,' where the model emphasizes proximity and co-occurrence statistics from the pretraining data, which can bias the model towards incorrect associations, causing hallucinations. 'Knowledge recall failures' also contribute, particularly with the inability to recall long-tail knowledge, which refers to the difficulty LLMs face in using seldom-occurring factual knowledge from the training data.

Another significant cause of hallucinations is linked to 'misuse of data and parametric knowledge' in complex scenarios, such as multi-hop question answering, where the reasoning engine must use multiple entity associations to respond accurately. This complexity often exceeds the LLMs' reasoning capabilities as established during the retrieval phase, leading to errors.

During the training phase, several factors induce hallucinations. 'Architectural flaws' related to the unidirectional representations in causal language models during pre-training can restrict the model's comprehension, as crucial context may appear beyond the immediate left-to-right scope. 'Attention glitches,' where the model fails to maintain attention over long sequences, can also result in hallucinations, as relevant information far from the current input token gets overlooked, particularly affecting long-term dependencies more than short-term ones.

'Exposure bias' during training introduces discrepancies between the training and inference phases in auto-regressive language models, leading to cascade errors during token generation due to reliance on previously generated tokens rather than ground truth tokens, which the model uses during training. Hallucinations during the alignment stage can also occur due to poorly formulated prompts and 'belief misalignment,' where the model produces content that does not align with the factual knowledge but rather follows the annotators' opinions during the RLHF process, a phenomenon known as sycophancy.

In the inference phase, decoding strategies inherently involve randomization, which introduces risks of hallucinations. The 'likelihood trap,' where highly probable sentences are not necessarily useful, is addressed by introducing randomness during decoding to create a more uniform token probability distribution, reducing the chance of sampling less frequent but contextually inappropriate tokens. The relationship between beam search, a decoding strategy that conditions sampling to promising tokens, and hallucinations is less explored but shows potential in reducing hallucination risks by limiting the sample space to tokens strongly connected to a likely token. Other causes of hallucinations during inference include 'imperfect decoding representation' and 'insufficient context attention,' which prioritize text fluidity over faithfulness due to top-layer transformer limitations, including the 'softmax bottleneck,' which restricts expressiveness.

10.3 Mitigation of hallucinations in LLMs

Detecting false factual information in LLMs is a challenging task. There are primarily two mitigation strategies: one involves automatic claim verification, and the other is based on estimating the uncertainty surrounding the model's output. Each strategy presents significant challenges: the first requires access to sources for verification, while the latter depends on interpreting the model's provided uncertainty value.

Concerning the use of external facts for verification, these strategies typically compare the facts generated by the model with those retrieved from knowledge bases. However, this approach is limited by the update lags in knowledge bases such as DBPedia. To address the issue of aligning results in time-sensitive contexts, it is necessary to resort to uncurated sources like web resources, which are inherently biased in various aspects. Ricardo Baeza-Yates identifies [16], in addition to conventional sources of bias such as sampling and algorithmic bias, biases triggered by user interactions on the web, such as self-selection and activity bias. As users influence the algorithms of personalized web recommendations through their clicks, their preferences reflect this self-selection bias in the content they see online. Furthermore, based on web interactions, which are dominated by a few, there is an activity bias that also influences recommendations. Baeza-Yates points out a vicious cycle of bias on the web. This cycle affects the web as a data source, making it a dubious choice for mitigating disinformation, suffering from the same issues as LLMs: unreliable, biased, and dominated by a few. The mitigation strategies for hallucinations, in this case, false factual information in LLMs, tend to rely on calculating trust indicators for the web sources used, an approach that mimics human efforts in various fact-checking initiatives.

Regarding uncertainty estimation as a disinformation mitigation technique, these strategies estimate the uncertainty of the factual content generated by the model. This can be done by examining the LLM's internal states, either through token probabilities or entropy. For tokens, a measure of uncertainty is the minimal token probability. The underlying assumption is that low probabilities indicate model uncertainty. In this line, the LLM's output can be used in a new prompt, instructing the LLM to generate a new output based on the previous one. The probabilities of the generated tokens will measure the LLM's familiarity with the generated factual knowledge and thus help us discard unreliable factual claims. Unfortunately, these strategies work in Open LLMs, such as LLama 3, and not in LLMs that can only be accessed through API calls, like those from OpenAI, since these do not provide the token probabilities of the output. Due to these restrictions, several studies have tackled the problem of uncertainty estimation from the perspective of LLM behavior. One way to do this is to formulate the prompt indirectly several times to the LLM, evaluating the consistency of the responses as a proxy for model uncertainty. These strategies depend on the methods used to formulate the indirect queries to the model.

Another dimension of analyzing the results of a LLM involves the detection of faithfulness. Faithful content can be identified using faithfulness classifiers. In this development line, there are methods based on textual entailment, which seek consistency between two consecutive sentences. In NLP, textual entailment is the task that helps determine if two consecutive sentences are linked by a premise-hypothesis relationship, meaning if one (the hypothesis) is a logical consequence of the other (the premise). In faithfulness detection, the output of an LLM can be broken down into sentences, and then textual entailment classifiers are applied to compute a faithfulness score.

Another line of analysis involves the use of question-answering based metrics, which are significant in NLP. These strategies involve identifying claims in a LLM output, then generating questions aligned with these claims. The questions are reformulated to generate answers and compare them with the original output's claims. By comparing the matching scores between the claims and the answers, we can calculate the faithfulness of an LLM. The limitations of these strategies are based on aspects like claim selection, question generation, and answer overlap, all of which have strengths and weaknesses that affect the reliability of faithfulness score estimations.

Similarly, the use of prompt-based metrics has gained attention. This very recent line is called LLM-based evaluation. The idea is to provide the LLM with clear instructions on how to evaluate certain tasks, so that the model itself assesses faithfulness. Different ways to evaluate the prompts include the use of chain-of-thought or allowing the model to generate evaluations along with explanations, all aimed at using either the chain of thought (a sequence of logical steps taken to generate the result) or the explanation as evidence to assess the faithfulness of the LLM.

10.4 LLMs imitating humans

The vast capabilities of LLMs to mimic human language represent a significant advancement in AI. LLMs effectively handle tasks such as machine translation, writing, paraphrasing, and various text production activities. They can tailor the text they generate based on defined writer profiles, exhibiting skills related to the imitation of colloquial language. These advancements are a major achievement for AI.

However, as observed, LLMs can generate "hallucinations," which affects the reliability of their results. Additionally, they can reproduce biases based on the parametric knowledge encoded in their foundational models, which often pull data from sources with varying reliability levels. The latest versions of ChatGPT include plugins to extract facts from external web resources. As discussed, the web has biases, and many of its sources are unreliable, thus heightening the risk of extracting false factual information from these sources. This makes LLMs unreliable in terms of generating factual knowledge.

Despite these issues, the potential misuse of these technologies is vast. According to Ferrara [100], the nefarious applications of LLMs produce various types of harm, including personal loss and identity theft, financial and economic damage, and information manipulation (see the mindmap in Figure 10.1). The main threats include:

1. **Personal Loss and Identity Theft:** GenAI has the potential to generate synthetic identities. Using text generation, user profiles can be written to create a false history. Fake profiles are the first step towards creating scams. As discussed later in the book, visual transformers and other extensions of these architectures have enabled the application of generative technologies to images and videos recently. This supports the creation of fake profiles with hyper-realistic images. The ease of use of these tools for generating manipulated content opens the door to digital impersonations, using a known personality via GenAI to impersonate people. These technologies can also be used in telephone scams through voice impersonation.
2. **Financial and Economic Damages:** The vast capability to generate unreliable content in large volumes can arm bad actors, who may flood social media with biased information. This can trigger alarms in the financial realm, even causing stock market crashes and allowing market manipulation.
3. **Information Manipulation:** The vast capability to generate unreliable content in large volumes can arm bad actors, who may coordinate influence campaigns on public opinion, flooding networks with bots and causing informational chaos. The use of propaganda strategies such as repetition, appeals to hatred, the black-white fallacy, and other persuasive linguistic strategies are in the hands of actors who can use them to sway public opinion, generating polarization and constructing controversies, with significant control over information on social networks.

One way to mitigate the nefarious effects of LLM misuse is to detect LLM-generated text. However, LLMs have advanced so much in text production that they are undetectable by humans. The famous machine intelligence test has long been the Turing test, based on the imitation game principle. As stated by Alan Turing, if two agents say a human and a machine, produce texts, and a human is unable to distinguish which is the human and which is the AI system, it indicates that the AI has managed to mimic human language and thus has generative text capabilities similar to humans. The imitation game, known as the Turing Test, has been surpassed by Chat-GPT, showing that humans are unable to distinguish between synthetic and natural text [32]. This mimetic capability leads to difficulties in detecting text generated by LLMs.

One approach to detection involves the use of algorithmic methods. Generating datasets for detection is complex due to the difficulty humans have

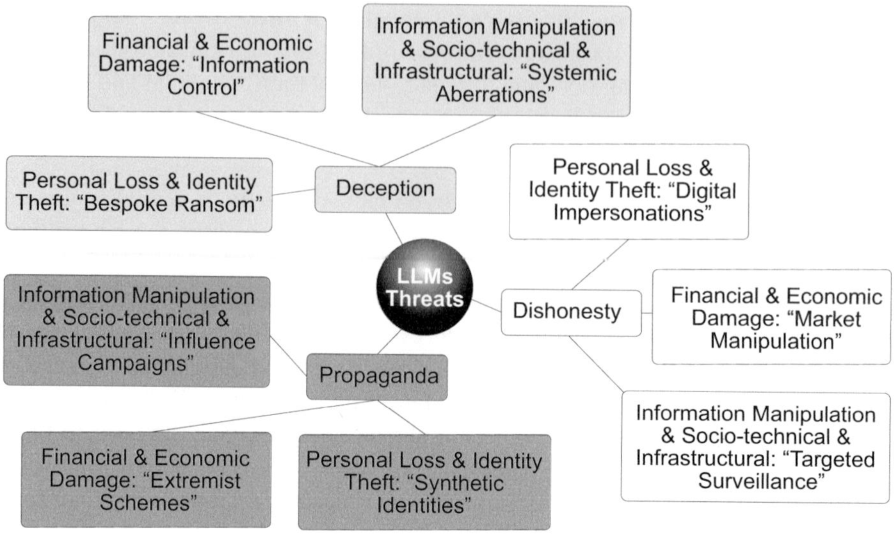

Figure 10.1: Some of the threats raised by LLMs mimicking humans identified in Ferrara [100].

in distinguishing and, therefore, annotating and supervising the creation of datasets for this task. However, there are some clues that can be focused on to discern these differences. OpenAI has released a classifier trained to distinguish between human-generated and synthetic texts [231]. This binary classifier was developed based on annotated texts and, while it is an initial advancement made by the developers of ChatGPT, it has limitations. Notably, it struggles with classifying short texts, making the detection of synthetic text on social media platforms unreliable. Another limitation relates to the overestimation of synthetic text; the classifier tends to misclassify texts actually generated by humans as AI-generated. Currently, it is only available in English. It also cannot distinguish synthetic text containing factually verified information, as the classifier confuses reliable output with human/artificial output. Another classifier designed for this purpose is GPTZero [122], which is trained to detect texts generated by ChatGPT, GPT-4, Bard, LLaMa, and other AI models. Although proprietary, this model provides various indicators that highlight clues distinguishing synthetic texts from human-generated texts. In Figure 10.2, we can see a comparison based on texts generated using ChatGPT on a foundational model GPT 3.5 versus texts collected from Twitter conversations on news sites. The Twitter16 dataset is relatively old, suggesting a low presence of bots, which allows attributing a good portion of these texts to humans initially.

The figure displays three metrics computed by GPTZero: readability, perplexity, and burstiness. Readability refers to how easily a text can be read by a user. In this metric, synthetic texts are slightly more readable than human texts, likely due to the more grammatically accurate constructions used by

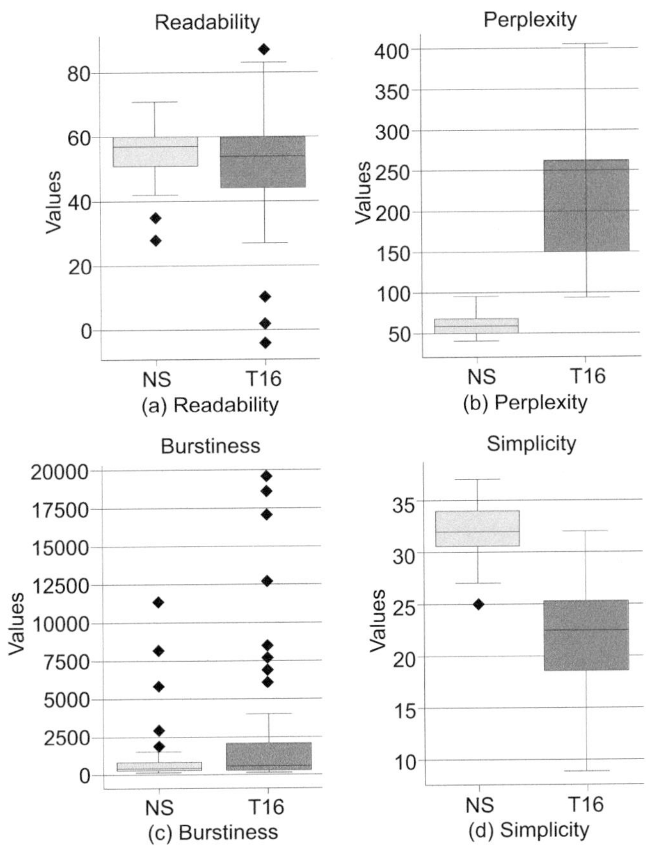

Figure 10.2: A comparison between natural conversations (T16) and synthetic conversations (NS).

LLMs, compared to the often confusing and convoluted grammar used by humans, especially on social media. Perplexity measures the predictive complexity of a text-based on autoregressive generation mechanisms. Clearly, the perplexity of synthetic text is lower than that of human text, indicating that human text is much more unpredictable than that produced by LLMs. This difference is due to the sampling strategies employed by LLMs, which rely on variants of conditioned random sampling, such as beam search. Humans are far less predictable in this regard. Burstiness assesses the repetitive use of the same tokens, a linguistic effect generally attributed to the use of slogans or other propaganda strategies. The figure shows that the differences in this characteristic between the two types of texts are very small. In summary, the comparison indicates that the main differences between human and synthetic text stem from the way LLMs sample vocabulary to produce their texts, with algorithmically generated texts being much more predictable than human texts.

Other methods to detect synthetic text include the use of algorithmic watermarking. The implementation of these strategies depends on their adoption by major tech companies, such as OpenAI.

10.5 Reflections on the benefits and risks of LLMs

LLMs offer significant benefits and opportunities. Their immense capabilities in text production can be advantageous if the technology is used for good purposes. However, a powerful technology like ChatGPT also carries risks of misuse. In the wrong hands, this technology can cause harm in various areas. The damage is primarily due to its ability to mimic human language, making it a tool with potential for misuse.

Current mitigation efforts rely on detection strategies, which currently have many limitations. These strategies require the development of specific AI models to detect other AIs, such as GPTZero. However, as detection technologies improve, so do the technologies for making outputs appear more human-like. As detection clues become clear, it is possible to algorithmically mask these characteristics, which makes the problem increasingly difficult.

In the following chapter, we will explore how extending generative models to images further extends the risks associated with GenAI. We will conclude that providing greater transparency to these models is essential and recommend some measures to address the inherent risks of these technologies.

10.6 Further readings

LLMs have raised significant concerns regarding safety and security risks, including ethical considerations, hallucinations, and prompt injection issues [111]. In environments where safety is critical, such as healthcare and finance, applying LLMs can result in model hallucinations, which may cause harm to vulnerable users [322]. Additionally, the threat of retrieval poisoning in LLM-powered applications allows malicious actors to manipulate the output, demonstrating a high success rate in real-world scenarios [3]. Integrating LLMs into educational settings also presents ethical challenges, such as data privacy concerns, bias, and the potential consequences of replacing human instructors [88].

Current research actively addresses these risks by examining the safety implications and ethical issues and identifying future research directions to develop safer and more ethical LLM applications [111]. One proposed solution involves leveraging a QA corpus to probe LLMs, manipulating both the prompt and knowledge representation to enhance accuracy in safety-critical environments [322].

Chapter 11

Visual Transformers and the Rise of Multimodality

11.1 Introduction

Generative AI has progressed towards handling formats beyond text. Perhaps the format in which it has shown the most surprising results is images. In just over a decade, the synthesis of images conditioned on a class, and later conditioned on a prompt, has advanced in unexpected ways. From its beginnings with Generative Adversarial Networks (GAN) [119] to the Visual Transformer (ViT) [84], the focus has been on generating high-resolution images. While most commercial models based on these architectures focus on generating images in various styles, generating hyper-realistic images is probably the development that has caused the greatest stir.

In this chapter, we will describe the rapid advancement of these technologies, as well as their applications and the main threats posed by their potential misuse.

11.2 Generative adversarial networks

The synthesis of images has been one of the primary objectives of AI in the last decade. One of the first successful architectures in this field was based on GANs [119], an artificial neural network architecture designed to estimate the parameters of a generative model using an adversarial process in which two models are simultaneously trained. One of these is a generative model that captures the distribution of the training data, which we shall refer to as model G, and a discriminative model D that estimates the probability that a sample originates from the training data. Both models, G and D, compete based on their objectives, defining an adversarial game that serves as a training strategy. While D attempts to minimize the probability of making errors, G maximizes the probability that D makes an error. In game theory, this type of competition is known as a minimax two-player game. Goodfellow et al. [119] demonstrate that in the space of arbitrary functions G and D, there is a unique solution such that G approximates the training data, and D is unable to distinguish between synthetic data and real data. That is, D, in the real/fake binary classification problem (real indicating that the sample originates from the data, fake indicating that the sample is synthetic), yields a $p =$ for any sample.

The initial GANs were trained using multilayer perceptrons for convenience, as they allow the application of the backpropagation algorithm to infer the parameters of G and D. Although jointly training both models entails computational challenges, Goodfellow et al. [119], from the University of Toronto, employed an alternating strategy, training D for k steps while keeping G frozen, and then freezing D and training G for one step (the term "frozen" indicates that the parameters will retain their values unchanged). Let p_g be the non-uniform distribution on transformed samples (synthetic) and let p_{data} be the distribution of the training data. It can be shown that the minimax game has a global optimum when $p_g = p_{data}$, and in this case, $D(x) =$ for all x. In practice, to produce transformed (synthetic) examples that allow G to be trained, denoted by the variable z, sampling is done from a noise prior $p_g(z)$. If G and D have sufficient capacity, both to approximate the data and to discriminate, then p_g converges to p_{data}.

In practice, GAN networks suffer from some issues, as convergence to the global optimum is not guaranteed because the convergence theorem depends on the capacity of G. Therefore, depending on the data, a multilayer perceptron might introduce several critical points in the parameter space, making it difficult to reach the global optimum. The early experiments based on GAN networks were used to approximate handwritten digits (a classic AI dataset known as MNIST), faces (Toronto Face Database), and objects or animals in the CIFAR 10 dataset (6000 images corresponding to aeroplanes, cars, birds, cats, deer, dogs, frogs, horses, ships, and trucks). While these early GAN experiments showed improvements in terms of metrics (results were reported based on likelihood estimates), human perception of these images was still far from being considered satisfactory. Although the images generated for MNIST and the Toronto Face Database (TFD) were quite convincing, the results for CIFAR displayed images with poorly defined colors and shapes. Both MNIST and TFD are datasets with less diversity than CIFAR, either due to the handling of color (greyscale) or by operating in specific domains (digits or faces). Some tests in these early GANs replaced the multilayer perceptrons with convolutional networks, an architecture specifically designed for 2D data (such as images).

The use of convolutional networks did not result in significant improvements in the quality of the generated images.

One factor that contributes to the difficulty in finding the global optimum, and thus leads to instability during training, is the use of large-scale models. While increasing parametric complexity enhances the capacity of these models, it also results in the emergence of more critical points in the parameter space. In machine learning, it is common practice to introduce a factor during training that penalizes the excessive use of parameters in each learning step. This penalization factor typically prevents models from overfitting and is generally referred to as a regularization factor.

In GANs, Brock et al. [41] from Google DeepMind explored the use of regularization in the generator (model G) by applying a strategy known as the truncation trick. This strategy allows them to exert greater control over the trade-off between sampling fidelity and diversity by reducing the variance of G's input. The use of this regularizer enables the model to scale up, increasing the parametric complexity of the generative model, thereby allowing it to operate on larger datasets such as ImageNet. This model, named BigGAN, shows significant improvements in the quality of synthesized images.

To manage the trade-off between fidelity and diversity, BigGAN introduces the truncation trick. This strategy involves modifying the prior $p(z)$ used for sampling z. The original GAN samples the z (the synthetic data) from a Gaussian distribution (specifically, z is drawn from $N(0,1)$). BigGAN samples from a truncated normal distribution, where values falling outside a certain range are re-sampled until they fall within the controlled range. It is important to note that the latent space is multidimensional, meaning that the z samples correspond to stochastic vectors. Therefore, strictly speaking, the truncation trick truncates the z vector by resampling the values whose magnitude exceeds a certain threshold. In the limit, when the truncation trick threshold approaches zero, the individual samples approximate the mode of G's output distribution. This technique allows them to manage a hyperparameter (the threshold), which can be calibrated post hoc based on metrics for synthetic image quality, such as the Frechet Inception Distance (FID). FID penalizes the loss of diversity, analogous to the recall in the precision/recall trade-off. However, FID also rewards gains in fidelity, analogous to the precision in the precision/recall trade-off. To ensure that this strategy yields good results, G is designed to be smooth so that the full space of z produces good output samples. To achieve this, Brock et al. use a variant of Orthogonal Regularization. With this regularizer, BigGAN was able to handle larger models that were compatible with truncation and could be calibrated based on FID by adjusting the threshold of the truncation trick. The results obtained by BigGAN on ImageNet were remarkable, not only for their improvement in evaluation metrics, primarily FID in this case, but also for the resolution and quality of the synthesized images.

This leap in quality is likely the first to bring synthetic image generation close to hyperrealism.

11.3 Diffusion models

Dhariwal and Nichola from OpenAI [79] demonstrated that diffusion models achieve better results than GANs. This seminal work opens a new line of research into these models, which dominated synthetic image generation for several years.

Diffusion models are a class of likelihood-based models that offer desirable properties such as distribution coverage, a stationary training objective, and scalability [141]. It is worth noting that GANs exhibited several instability issues during training, primarily triggered by their lack of scalability. Even though BigGAN addressed these challenges, problems with instability and difficulties in handling critical points during training persisted, making the training of such networks quite complex. Diffusion models, on the other hand, tackle these challenges through a different approach. These models generate samples by gradually removing noise from the signal (the data), and their training objective is expressed as a reweighted variational lower-bound. Inspired by the truncation trick, Dhariwal and Nichola [79] bring the benefits of addressing the diversity-fidelity trade-off in diffusion models.

In simple terms, diffusion models use gradual steps to obtain samples that progressively contain less noise until, eventually, they achieve a sample from the training dataset. For this, at each time step, a noise level is considered, which involves a mixture of the real data and noise, where the signal-to-noise ratio is determined at the timestep. Typically, the noise is Gaussian. The chain of samples obtained by controlling the signal-to-noise ratio allows for progressively denoised samples of the original data. The model can be parameterized as a function that predicts the noise component of a sample, meaning a loss function is defined that corresponds to the mean squared error between the true noise and the predicted noise. To sample from the noise predictor, the distribution $p_\theta(x_{t-1} \mid x_t)$ is modelled as a Gaussian whose mean is calculated from the noise predictor and whose variance is fixed (the Gaussian is diagonal, so the covariance matrix is diagonal). The entire parameterization of the diffusion model was originally introduced by Ho et al., who proved that this approach allows for high-quality samples when the total number of diffusion steps is sufficiently large.

The architecture used by Ho et al. [141] is U-Net. U-Net is a type of convolutional neural network (CNN) specifically designed for biomedical image segmentation tasks. Initially developed by Olaf Ronnenberger et al., Philipp Fischer, and Thomas Brox in 2015 [257], the U-Net architecture is characterized by its symmetric, "U"-shaped design, which enables the efficient transmission of contextual information through the network's layers. This architecture consists of two main components: an encoder that captures the image's context and a decoder that facilitates precise localization. The skip connections between the encoder and decoder are crucial, as they combine high and low-resolution features to enhance segmentation accuracy.

From a technical perspective, the U-Net encoder comprises convolutional blocks followed by max-pooling operations to reduce spatial dimensions and increase the depth of the extracted features. In contrast, the decoder utilizes upsampling operations and convolutions to reconstruct the output to the original image resolution. The skip connections between corresponding layers of the encoder and decoder help to preserve fine details in the segmented image, thus overcoming typical challenges of information loss in deep networks associated with the vanishing gradient problem.

Dhariwal and Nichol [79] extended the capabilities of U-Net by incorporating attention layers at various resolutions and increasing the number of attention heads. One can conclude that Dhariwal and Nichol's efforts aimed to align the U-Net architecture more closely with the Transformer architecture by integrating several key elements from it.

Another significant aspect of Dhariwal and Nichol's work involved exploring the strategy of conditional generation, which they refer to as classifier guidance. GANs utilize class labels during generation, defining the generation problem as a class-conditional one where discriminators are explicitly designed to function as classifiers. In the diffusion model, a classifier is trained on noisy images, and the classifier's gradients are used to guide the diffusion sampling process towards an arbitrary class. This strategy resembles the discriminator's role in GANs, where the classifier aims to distinguish between real and fake samples. In diffusion models, the gradient of a pre-trained sample classifier is used to guide the diffusion sampling process towards arbitrary classes. This is known as a conditional reverse noising process.

To apply classifier guidance to a large-scale generative task, the authors trained classifiers on ImageNet. The classifier is based on the same UNet architecture, modified for stable diffusion, focusing on a specific layer (the 8×8 resolution layer), which produces the final output. The classifiers are trained on the same denoising distribution used by the diffusion model. After training the model, the classifier is incorporated into the diffusion sampling process.

One technical challenge encountered was that using gradients without scaling did not allow the classifier to be trained adequately. However, by scaling the gradients, the classifier achieved near-optimal results. Experiments on ImageNet demonstrate that the combination of these factors—conditional generation, class guidance, and gradient scaling (by a factor of 10)—enabled the authors to achieve what were at that time state-of-the-art results on a metric known as the Inception Score (IS). Conversely, the effect of scaling was less significant when evaluating based on the FID score, meaning the best results were obtained using only conditional generation and class guidance. This outcome is related to the fact that FID requires optimizing both diversity and fidelity, whereas IS only measures fidelity. The experiments also showed that class guidance is superior to BigGAN in addressing the diversity-fidelity tradeoff.

From a human perception perspective, samples generated by BigGAN exhibit distortions in markers of realism, such as facial geometry. In contrast, diffusion models show greater consistency with physical constraints, both in handling shapes and colors, thus approaching hyperrealistic image synthesiz much more closely.

One limitation of this model is that class guidance, a key element in generating hyperrealistic images, relies on the existence of labelled data, making it less suitable for unsupervised scenarios. A potential solution anticipated by the authors was to generate synthetic labels based on clustering samples.

11.4 Image generation conditioned on text

A natural progression in synthetic image models is the development of image generators conditioned on a prompt. In this context, Radford et al. at OpenAI [241] tackled this challenge by developing the CLIP model (Contrastive Language–Image Pre-training). The key to CLIP lies in working with aligned texts and images, i.e., a descriptive text of an image (image caption), training a model that predicts correct pairings of a batch of (image, text). This pairing model is pre-trained to be subsequently used in transferable visual models.

To train the pairing matcher, multiple texts and images are used, aligning the image with its caption based on the maximization of the product of the text and image encodings. Once the text-image pairing is trained, a dataset is created from the caption by replacing a keyword with alternative options (mimicking what is done in masked language models). For instance, if the image caption is "A photo of a dog," the masked language model strategy produces an example like "A photo of a object," where object can be instantiated in various ways (e.g., plane, car, dog, ... bird). This masking substitution strategy generates *N* samples from the image caption, of which only one is correct (dog). By using the text-image pairing, it is expected that the correct caption is produced, and the alternative captions are discarded based on the image encoder. This model operates in a "zero-shot prediction" mode, as it can be seen that we have transferred from the image encoder to the text space using unsupervised data or other types of explicit transfer strategies.

These types of models are known as weakly supervised models, because by using the image captioning trick, it is possible to have vast volumes of paired data using public sources such as Wikipedia. The creators of CLIP worked with a dataset of nearly 400 million image-caption pairs. Experimental results based on this model showed that, similar to GPT, CLIP learns a wide range of tasks, this time in a bimodal way, and is therefore useful for OCR, geo-localization, action and posture recognition, among other common tasks in the computer vision field. The transferability evaluation of CLIP across nearly 30 existing datasets in the computer vision field illustrated the flexibility of this model in different tasks, even showing its superiority over models specifically trained on each of these datasets.

Undoubtedly, CLIP constituted a fundamental advance in bimodal generative models, in this case, in the text-image modality. In terms of architecture, CLIP experimented with various encoders for both images and text. It was concluded that the transformer was the most suitable architecture for encoding text, while the ResNet architecture was used as the image encoder. ResNet (Residual Networks) is a deep neural network architecture introduced by He et al. in 2016 [137] that facilitates the training of significantly deeper networks for images by using skip connections. These connections allow input signals to flow directly to deeper layers, mitigating the vanishing gradient problem that often occurs in very deep networks. The main component of ResNet is the residual block, which integrates the block's input with the output of a series of transformations using a skip connection that adds the original input to the output of the intermediate convolutional layers. This architecture proved to be effective not only in improving accuracy in image classification and detection tasks but also in accelerating the training process of CLIP.

A second architecture utilized in CLIP for encoding images was the Visual Transformer, an architecture proposed by Dosovitskiy et al. [84]. The Visual Transformer (ViT) architecture marks a significant innovation in the field of computer vision by adapting transformer architecture to image analysis. ViTs operate by dividing an image into patches, which are then flattened and transformed into data sequences, similar to the way words are handled in text processing. These patches are processed through multiple transformer layers that employ attention mechanisms to capture global dependencies among them. Unlike convolutional architectures, such as ResNet, which focus on local areas, transformers have the capability to attend to any part of the image due to their ability to configure global attention strategies. This provides a more flexible and potentially more powerful approach for tasks such as image classification and object detection.

A distinctive feature of ViTs is their reliance on the attention mechanism, which allows the model to weigh different parts of the image based on their relevance to the task at hand. This is achieved by calculating attention scores that guide the model's focus towards the most significant interactions between patches. As transformers do not inductively incorporate spatial biases like CNNs, they require large amounts of data and computational power to train effectively. Given that CLIP was trained on an enormous dataset of 400 million image-text pairs, the use of ViT in training CLIP was appropriate.

The creators of CLIP reported no significant differences in terms of the time required to train CLIP. While the ResNet-based version took approximately 18 days on 592 V100 GPUs, the ViT-based version took 12 days on 256 V100 GPUs. This demonstrates that ViT offers advantages in training efficiency compared to ResNet. The reported results indicated that ViT achieved superior performance, leading to the final implementation of CLIP being based on the ViT architecture.

11.5 Diffusion models with transformers

The success of the ViT architecture during the development of CLIP drove the next significant technological advancement in image generative models: the development of diffusion models based on the transformer architecture. This advancement, pioneered by Peebles and Xie [237] (with Peebles conducting this work within the FAIR group at META AI), replaced the U-Net architecture previously used in stable diffusion models with the successful ViT architecture, which had already demonstrated promising results in CLIP. Essentially, the ideas of stable diffusion were implemented on ViT based on latent patches. This development demonstrated that the inductive bias of U-Net was not crucial for the performance of diffusion models and that it could be substituted with the transformer. This new model, which integrates diffusion models with transformers, was named Diffusion Transformers (DiT).

An additional modification introduced by DiT is that it does not operate directly on image patches, as the original ViT did. Instead, it adopts the strategy introduced by Rombach et al. [254], operating on patches from the latent space. These models function by reducing the dimensionality of images to a lower-dimensional latent space, where the diffusion process is carried out more efficiently. The diffusion model learns to gradually generate new samples from a noise distribution, iteratively refining these samples in the latent space before mapping them back to the high-resolution space—a strategy based on the Variational Autoencoder (VAE) [170].

Technically, the process begins with encoding high-resolution images into compressed latent representations using an autoencoder. The key to this approach is that, in the latent space, a diffusion model is applied, trained to model the distribution of the latent representations. This diffusion model gradually reverses a noise process added to the latent representations, allowing for the generation of new images by decoding the refined latent samples. This technique not only significantly reduces computational costs compared to diffusion models operating directly in high-resolution image space but also maintains or even improves the quality of the generated images.

The combination of both strategies—the transformer applied to patches and the diffusion process from the latent space—are two key factors that make DiT work. While the use of the transformer allows for high-resolution results, applying the diffusion process in the latent space enables DiT to scale in terms of the amount of data it can handle.

A recent advancement in this architecture involves incorporating a final modification to DiT models: rectified flow. Rectified flow is a generative formulation that connects data and noise in a straight line. The idea behind this strategy is to enhance the existing noise sampling techniques used during the diffusion process, biasing them towards perceptually relevant scales. Additionally, this model, referred to as the rectified flow transformer, modifies the DiT architecture for text-to-image synthesis by using separate weights for text and image, allowing for bidirectional information flow between images and tokens. [96] demonstrate that these modifications improve text understanding in images, typographic synthesis, all of which imply enhancements in human perception of the generated images.

Rectified flow transformers use CLIP-based encoders to handle texts. The architecture also adds a T5 encoder (text-to-text transformer encoder-decoder), forming a joint representation of the input caption data. On the other hand, the encodings obtained from CLIP are fed into a MLP and combined with an encoding representing a timestep. This enables the architecture, besides processing captions as sequences of symbols, to keep track of the order in which these symbols were presented to the transformer. The architecture also incorporates the noised latent for the image patches, which pass through a linear layer and are added with a positional encoding to record the position of the patch within the image. In summary, the architecture generates three inputs for the transformer blocks: those from the caption (processed by CLIP and T5), the timestep combined with the caption encoding extracted from CLIP, and the image patch encoding conditioned on the noised latent, which includes a positional embedding. These three inputs are fed into the transformer. While the image patch embedding and the caption embedding enter the transformer blocks, the caption embedding combined with the timestep encoding is used to feed the residual connections of the transformer. Thus, these skip connections enter at different levels of processing performed by each transformer block. Concerning the transformer blocks, they independently process the image patch embedding and the caption embedding, each entering a linear layer. These two inputs are combined to generate the QKV inputs used by the transformer's self-attention layer. The combination is performed through element-wise multiplication. From the attention module, two copies of the output encodings are extracted, which are fed into the next transformer block, repeating the process.

The architecture achieved state-of-the-art results on ImageNet using the FID metric. One insight derived from the evaluation of the architecture is its enhanced capabilities in handling captions. This improvement not only enhances the resolution of the generated images but also provides captions of higher quality, both in capture and synthesis. All these aspects improve the model's language understanding capabilities, enabling it to capture the prompt's content more effectively when generating new images. The results presented surpass those achieved by other models, even when considering commercial models like OpenAI's DALL-E, which are based on diffusion.

11.6 Multimodal integration in conversational models

The progress demonstrated by the integration of transformers with stable diffusion models has been so remarkable that it has spurred the development of models incorporating bimodal language understanding. Among commercial models, perhaps the most notable are those developed by OpenAI, with their flagship model being GPT-4 [230]. GPT-4 possesses the capability to simultaneously process text and images, opening up a vast range of applications.

GPT-4 was originally launched to process text and images as input and produce text as output. Its primary strength lies in generating natural language in more complex scenarios, with prompts that integrate both text and images. A surprising outcome is that GPT-4 not only exhibits the ability to handle bimodal data but also shows significant improvements in language understanding. This is due to a combination of factors from which GPT-4 benefited, namely, incorporating more data by using bimodal data and also increasing the model size.

Regarding model size, these models have grown immensely. While GPT-2 in 2019 had 1.5 billion parameters, GPT-3 in 2020 increased to 175 billion parameters. GPT-3 and its 2022 version, GPT-3.5, form the basis of ChatGPT, released in November 2022. From these versions, which operated around the order of 175 billion parameters, GPT-4, released in March 2023, grew to nearly 1000 billion parameters. The efficient version of GPT-4, GPT-4 Turbo, was released in November 2023. It is known that GPT-4 is a larger model in terms of parameter capacity and computational requirements, allowing it to understand and generate texts with a higher degree of complexity and fidelity. On the other hand, GPT-4 Turbo is optimized to offer faster responses and more efficient use of computational resources, making it a more agile and economically accessible version, but potentially with fewer capabilities in tasks involving text generation and contextual understanding compared to GPT-4. This "Turbo" version is designed for applications requiring low latency and rapid performance, ideal for environments where response speed is critical. It is believed that the Turbo version has, therefore, fewer parameters than the GPT-4 version released in March 2023.

In 2024, conversational systems have again brought many surprises. The multimodal integration, already explored in GPT-4 based on the ViT architecture, expanded to include other modalities, specifically audio. The use of speech understanding models represents a significant advancement in these types of systems. This had already been explored in various voice command systems, notably in Alexa (Amazon) and Siri (Apple) assistants. At the heart of these systems are Automatic Speech Recognition (ASR) models, which convert human voice audio signals into text. Modern ASR systems typically employ artificial neural networks to capture the temporal dependencies and spatial features of audio. Once the audio is converted to text, LLMs are used for language understanding tasks. The union of both technologies, speech-to-text and LLMs, is a fruitful combination. In fact, developers of LLMs have incorporated this technology to facilitate interaction with humans. For instance, OpenAI has developed a highly successful speech-to-text model named Whisper. The Whisper model is an Automatic Speech Recognition (ASR) system based on the transformer architecture. This model is notable for its ability to transcribe text from audio in multiple languages and dialects with high accuracy. It uses a full attention transformer, which processes audio sequences to predict corresponding text transcriptions. The key to its effectiveness lies in its training, which was conducted using a massive and diverse dataset encompassing a variety of languages, accents, and acoustic conditions. This approach enables the model to effectively handle linguistic and acoustic variations, making it a robust tool for transcription applications in different contexts and environments.

Whisper implements a token-based encoding approach that transforms audio inputs into latent representations, which are then processed by transformer blocks to predict text sequences. Additionally, the model benefits from an attention mechanism that allows it to focus on specific parts of the audio input, thereby enhancing transcription accuracy. The practical implementation of Whisper also includes features such as automatic language detection and the ability to handle low-quality audio. This flexibility has driven the development of OpenAI's latest model, GPT-4o (the "o" standing for "omni"), an end-to-end model that incorporates image, audio, and language understanding, building upon GPT-4. GPT-4o was launched in May 2024 and is available for desktop versions, API access, and mobile access.

META has not lagged behind. Through its LLMs, LLama has taken a significant position in the realm of LLMs. A major advantage that META holds over OpenAI is its operation on social media platforms. Given that META manages platforms such as Facebook, Instagram, and WhatsApp, the integration of Llama models into these platforms is a natural progression. Indeed, Meta has activated its LLM in a WhatsApp account named Meta AI. Meta AI operates on the Llama 3 LLM and can be activated in our WhatsApp application as an additional contact in our contact list. The activation of Meta AI in WhatsApp has been available since July 2024 on all our mobile phones.

Another major technology company that has made significant efforts to keep pace in the LLMs arena is Google. Google has been a key player in the development of information technologies over the past two decades. Since its inception, its flagship product has been its search engine. Google's search engine led the Web 1.0 era, being the most powerful and widely adopted engine in the West. In a highly dynamic environment, the emergence of social networks spurred new business opportunities in tech, and other players entered the tech domain in the second decade of the 21st century. The rise of Facebook and later Instagram, now under META, has complicated the tech landscape. This intense competition among these giants has spurred further advancements and triggered significant investment in technology development. The dominance of AI has undoubtedly been the cornerstone of this race over the past five years. In this context, Google had already developed its Google AI environment, providing AI tools for developers. Google's relevance in NLP was fundamental in machine translation systems, and Google Translate is likely the first NLP system to achieve widespread adoption globally. However, its entry into the race for LLMs was somewhat delayed. Considering that the first GPTs appeared shortly after the introduction of the Transformer architecture, which can be dated to 2018, the release of Bard is indeed late. Google launched its AI model known as Bard in March 2023. Bard is based on the Language Model for Dialogue Applications (LaMDA) architecture, specifically designed to improve text generation in conversational contexts. This model utilizes deep learning techniques to understand and generate human language more naturally and contextually. LLM leaderboards, based on evaluations across various benchmarks such as MMLU (Massive Multitask Language Understanding), have consistently shown that Bard lags significantly behind OpenAI's models.

Google has made significant efforts to regain ground in this area, investing in its new conversational model, Gemini. Launched on December 6, 2023. This model represents a major advancement in the field of AI language models. Developed as a response to the progress made by generative models such as GPT, Gemini stands out due to its bidirectional attention architecture and its efficient training capabilities across a variety of linguistic tasks and multimodal data processing. This model is built on advanced deep learning techniques, including optimizing transformer architecture and federated learning techniques.

Given its relevance to Gemini's training, it is important to briefly explain federated learning. Federated learning is a machine learning paradigm that enables the distributed training of AI models while preserving data privacy. In this approach, multiple servers participate in training a global model without sharing their local data. Each node trains a copy of the model using its own dataset and then sends only the model parameter updates to a central server. This server aggregates the updates to improve the global model, which is subsequently distributed to all nodes for further iterations. This method not only helps protect privacy but also reduces the need to transmit large volumes of data.

Another key player in conversational AI is Microsoft. Its strong interest, coupled with close collaboration with OpenAI, has allowed it to gain ground in the race for LLMs. Its Azure environment is undoubtedly its flagship product for solution developers, prominently featuring embedded access to GPT-4. By 2023, this alliance had already begun to bear fruit with the integration of LLMs into Azure, and subsequently into several of Microsoft's flagship products, such as its Bing search engine. The integration of LLMs into Bing's search engine warrants special attention. Bing has integrated advanced AI technologies, particularly OpenAI's GPT-4, to enhance its search and response capabilities. This integration is achieved by connecting the search engine with an AI API, allowing Bing to process natural language queries and generate responses that are not only relevant but also contextually accurate. AI models trained on extensive datasets can understand and generate natural language, facilitating a more intuitive and efficient interaction with users. Additionally, the integration of continuous machine learning capabilities allows Bing to adapt and improve its responses based on user interactions and changes in available information on the web. The integration of Bing's search engine capabilities with GPT-4's language understanding is undoubtedly the most notable aspect of this partnership between Microsoft and OpenAI.

Microsoft has also explored the development of LLMs. Its flagship models, known as Phi models, are particularly notable for being smaller in terms of the number of parameters. Microsoft has developed models that are smaller than OpenAI's, primarily due to efficient tokenization and parameter sharing, which reduces redundancy while maintaining or even enhancing model performance in NLP tasks. Furthermore, Microsoft incorporates quantization and pruning methods during and after training, resulting in lighter and faster models. These approaches enable Phi models to handle large volumes of data and complex inference operations more efficiently, which is crucial for real-time applications and devices with limited resources, such as mobile devices.

11.7 New players enter the field of conversational AI

An interesting development in the race of LLMs is the emergence of players who are outsiders to the major corporations. In fact, even OpenAI can be considered an outsider, with origins that were initially closer to an academic initiative than to a large company. In this context, companies like Anthropic and Perplexity are particularly noteworthy for study.

Anthropic is an AI company founded in 2021 by former members of the OpenAI team, including Dario Amodei, who was Vice President of Research at OpenAI. The company was established with a focus on developing safer and more ethical AI technologies, aiming to address fundamental issues related to the alignment and governance of large-scale AI models. Anthropic's early products include Claude, a large-scale language model designed to be interpretable and less prone to generating harmful or misleading content. This model was developed as an alternative to more well-known models like GPT-3, with specific improvements in robustness and safety through techniques such as causal interpretation and constitutive correction in model training.

Recently, Anthropic has released the Claude 3.5 Sonnet model, an LLM with substantial language understanding capabilities, approaching the results that can be achieved using GPT 4. This model is part of the ongoing iteration to enhance the safety and alignment of large-scale language models, incorporating innovations in training and model design. Building on previous advancements of Claude, this model integrates specific techniques to mitigate risks such as the generation of false information and the strengthening of filtering algorithms to avoid undesirable content. Through an iterative and supervised training approach, Claude 3.5 Sonnet improves on key aspects such as context comprehension and the generation of more coherent and contextually appropriate responses. Its development marks a significant milestone in the pursuit of AI systems that are not only powerful in linguistic capabilities but also safer and more aligned with ethical and social expectations.

Perplexity.ai, of more recent origin, was launched in 2023 as a start-up in AI, focused on the development and commercialization of advanced language models. Perplexity.ai is closer to what we know as a search engine integrated with a LLM, similar to Microsoft's integration of Bing with GPT. Perplexity places greater emphasis on the handling of URLs, resembling more closely what is observed in a traditional search engine. It explicitly declares these sources in its responses, thereby contributing to transparency and data traceability, something that is generally challenging to achieve in other LLMs.

11.8 Hyperrealism in motion

Recent advancements in generative AI have enabled the creation of hyperrealistic moving images. While the technical specifics behind these impressive developments remain largely unknown, there are indications that models like Sora for video creation from text may also rely on ViS or DiS architectures. However, instead of operating on image patches, they work on sequences of patches, where the sequence exhibits temporal interdependence. This is discussed in the technical report available at `https://openai.com/index/video-generation-models-as-world-simulators/`, which inclu- des examples of extraordinary realism, representing the next wave of models—this time featuring moving images.

Despite the highly restricted access to models like Sora and the prohibitive costs associated with replicating a model akin to OpenAI's achievement, which create a significant gap in technological reproducibility, it nonetheless demonstrates that the boundaries of generative AI continue to be pushed forward. The primary concerns associated with these advances revolve around the significant gap in computational infrastructure necessary to train these models. This greatly hinders the development of open models, making it seem that such models are primarily focused on a very specific area, particularly within the creative industries such as film production or commercial advertising. Perhaps in the coming years, we will see technological barriers to developing these technologies lower, allowing for the availability of lighter video generation models that could lead to the widespread adoption of this cutting-edge technology. For the time being, however, it seems to be confined to a few leading-edge, predominantly commercial laboratories.

11.9 Who wins the race?

With so many participants in the race, it is difficult to predict what will happen in this field. The highly dynamic nature of this scenario makes it challenging to foresee the trajectory of AI in the coming years. Understanding how LLMs truly compare, their strengths and weaknesses, is crucial as we enter a process of ethical reflection, which is the focus of the conclusion of this book.

Determining which LLM is superior in a specific task or in understanding language within a particular domain is a challenging task that has the community closely monitoring the various initiatives emerging in this field. Essentially, there are different evaluation datasets designed to assess various capabilities of LLMs, typically based on multiple-choice questionnaires with four possible answers across different subjects. As previously mentioned, the MMLU is one of the most widely used questionnaire datasets for evaluating LLMs. It includes questionnaires across 57 different topics. These topics prominently feature disciplines such as Social Sciences (12), Humanities (13), STEM (19), and others (13). These disciplines cover knowledge across various educational levels, ranging from college, high school, to university levels. Being a dataset designed in the Northern Hemisphere and primarily based on questionnaires related to standardized tests, there is a mix of universal topics and others that reflect the local context. Among the latter, humanities-related knowledge is highlighted, such as US History, European History, World History—primarily Western—and social sciences like US foreign policy. Other areas with a more global reach mainly include STEM topics. The use of this resource in LLM evaluation, although it has universal coverage in certain areas, raises the discussion about the inclusion of topics with local relevance. Its impact on leaderboards and, therefore, the incentive for LLM developers to include these topics is evident, which we referred to as an evaluation bias in the earlier chapters of this book. The need for resources that incorporate greater cultural diversity, especially in the disciplines of social sciences and humanities, is urgent as it distorts the development and evaluation landscape of LLMs.

Other highly relevant resources that are widely used in LLM leaderboards include evaluation datasets in Mathematics (MATH Lvl 5), GPQA (advanced knowledge in basic sciences), and MMLU-PRO, a more sophisticated version of MMLU where the questionnaires include 10 alternatives instead of 4.

This is a highly relevant aspect of LLM evaluation centres on their reasoning capabilities. Notable among these are BBH (Big Bench Hard), which includes tasks related to algorithmic reasoning and world knowledge, and MuSR, a dataset for evaluating multistep reasoning (multiturn questions) that also involves complex algorithmic problems. The problems are primarily of a scenario-based nature and include murder mysteries, object placement questions, and team allocation optimizations.

Regarding leaderboards, several initiatives exist. One noteworthy example is the Open LLM Leaderboard, available on Huggingface, which focuses on non-commercial LLMs (see `https://huggingface.co/spaces/open-llm-leaderboard/open_llm_leaderboard`). Additionally, there are initiatives where users design questionnaires, and based on the responses, points are awarded to various LLMs. In this context, the Chatbot Arena leaderboard (`https://huggingface.co/spaces/lmsys/chatbot-arena-leaderboard`) stands out, featuring both open and commercial LLMs. As of August 12, 2024, this leaderboard had accumulated 1,671,145 votes based on evaluations of 128 models. On this date, the top spot in the arena score (the score calculated based on user evaluations) was held by ChatGPT-4o (version of August 8, 2024), closely followed by Google's (DeepMind) Gemini-1.5-Pro-Exp-0801. The third position, based on the arena score, was occupied by GPT-40 (version of May 13, 2024). Finally, the fourth position was shared by GPT-4o-mini, a lightweight version of GPT-4o in terms of parameters, alongside Claude 3.5 Sonnet by Anthropic, Gemini Advanced App by Google, and Llama-3.1-405b- Instruct by Meta. It is noteworthy that no open LLMs manage to compete with the major companies in the leaderboards. It is also interesting to note that among the companies leading the LLM race, two well-known players, Google and Meta, are prominent, while two new entrants, OpenAI and Anthropic, have emerged, both products of the LLM era. This is a tightly contested race with an uncertain outcome.

In relation to the visual language understanding capabilities discussed in this chapter, the same leaderboard shows that fewer models are competing. Chatbot Arena features 15 models evaluated based on 65,415 votes (as of August 11, 2024). Two multimodal models, Gemini-1.5-Pro-Exp-0801 and GPT-4o (version of May 13, 2024), lead the rankings. The third place is held by Claude 3.5 Sonnet. Overall, the dominance of proprietary models on this leaderboard is significant, with only a few open models appearing among the top ranks, such as the InternVL2-26b model by OpenGVLab, LLaVA-v1.6-34B by LLaVa, and

CogVLM2-llama3-chat-19b by Zhipu AI. The remaining models are all proprietary, indicating a close competition between Google's Gemini models, OpenAI's GPT, and Anthropic's Claude.

11.10 The risks of multimodal models

At present, with a clear view of how LLMs, whether conversational text-based or the recent ones with multimodal features, have dominated the most significant advances in AI in recent years, we can reflect on the risks involved in the development of these technologies.

Although the emergence of manipulated content predates the development of generative AI—primarily through technologies involving video or image alignment and fusion—the potential for the widespread dissemination of manipulated content such as deepfakes is now immense. Regarding deepfakes, it is well known that the earliest results based on technologies like face swapping used conventional computer vision algorithms, such as interest point detection. The process of generating such content was significantly more complex, requiring specific development and settings to create it. In other words, the original technology necessitated extensive editing work and a setup that involved, for example, recording content to impersonate voice or image for face swapping. Due to the complexity of the experimental setup, the content could be hyper-realistic, but generating a large volume of this type of content was very challenging.

What changes with generative AI is the ability to generate images at will based on prompts, or even videos, which significantly increases the potential to flood the information ecosystem with this content. Since the barriers to generating hyper-realistic images—and in the near future, videos—have been lowered, access to this technology is now more widespread, and the potential for misuse has increased. Some researchers, such as Emilio Ferrara, foresee a scenario conducive to identity theft and the widespread dissemination of misinformation on social networks [101]. We share these concerns and foresee a very complex scenario for Western democracies and free expression. It will be the responsibility not only of large companies but also of regulatory institutions and civil society to ensure that the enormous advances brought about by AI do not turn against us.

11.11 Further readings

Multimodal language models represent a significant advancement in AI, particularly in the generation and manipulation of multimodal image/text content [312]. These models have demonstrated exceptional capabilities in generating text that can, on occasion, surpass human performance, showcasing their potential to revolutionize industries such as finance, business, healthcare, and cybersecurity Their ability to comprehend and generate complex content holds considerable promise for enhancing AI applications across various domains. However, the deployment of these models is not without ethical concerns and risks [308]. Multimodal language models can inadvertently generate harmful content, including offensive texts and inappropriate images, which poses substantial ethical risks [323]. Other ethical challenges include the potential for these models to perpetuate biases and erode collective knowledge within the digital ecosystem, leading to significant social and epistemic dilemmas. Additionally, concerns about their trustworthiness and the potential safety and security risks they present underscore the crucial need for developing frameworks that prioritize fairness, transparency, explainability, and accountability [316].

To address these concerns, considerable efforts are underway to develop robust mechanisms to mitigate the risks associated with these technologies [263]. Innovative toxicity metrics, detoxification methods, and algorithms that align ethical values with language models are currently being researched and implemented [214]. Furthermore, fostering responsible innovation and creating a comprehensive taxonomy of the ethical and social risks are vital steps towards integrating multimodal language models into sensitive areas, including the medical field [182]. These efforts underscore the commitment to ensuring that the benefits of these technologies are realized while minimizing their potential harms [86].

Chapter 12

Perspectives and Challenges

12.1 Introduction

Throughout this book, we have explored fundamental principles of AI ethics, presenting key examples, issues, and definitions. Our objective has been to demonstrate not only the importance of these topics but also the intricate challenges involved in the ethical development of AI.

In this final chapter, we will address two additional aspects of this discussion. First, we aim to clarify what we believe constitutes the integration of ethics into AI, particularly through what we term an "ethical unveiling," which we argue is central to the key advancements in the emerging third wave of AI ethics. Second, we will briefly share our vision for the future of AI development in the coming years, identifying critical areas of concern that, in our view, must be carefully considered as the field evolves.

12.2 Ethical unveiling for the AI third wave

As mentioned in Chapter 3, recent research suggests that the third wave of AI development involves a shift toward considering sustainability and ecological impact. However, we also want to emphasize another significant challenge related to criticisms discussed in Chapter 1, where some have labelled AI ethics principles as "useless."

While the second wave of AI ethics focused on raising awareness about the importance of ethical considerations, the current challenge is translating that awareness into specific practices and methodologies that effectively integrate AI ethics into the design, development, and implementation phases. Johnson and Verdicchio [159] have provided conceptual guidance on understanding this integration. They criticize the simplistic notion that ethics can be directly "added" to AI to create what might be called "ethical AI," such as a machine learning system with embedded ethical values. In their analyses [160], they describe this belief as a "fallacy of addition," arguing that combining AI with ethics does not automatically produce an ethical system because the two elements are ontologically distinct. AI should be understood as a set of computational artifacts, while ethics is a set of human and social values that cannot simply be integrated into AI, as they belong to different ontological categories.

The authors explain that for ethics and AI to be effectively integrated, both must share ontological characteristics, implying that ethical principles would need to be computable. However, they argue that this is improbable because ethical values are inherently abstract, social, and contextual, while computational artifacts function based on concrete, deterministic processes, and data, often reducible to binary terms and numerical values. So, what options do we have?

As we have discussed throughout this book, we conceptualize AI as a sociotechnical system. This perspective enables us to explore ways to integrate ethics into AI without falling into the fallacy of treating ethics as an external component that can be simply "added" to AI.

When AI is viewed solely as a computational artifact, ethics and AI cannot be easily combined because they belong to different ontological categories. However, the relationship between the two becomes more nuanced and complex when one understands AI as a sociotechnical system—where AI and values coexist within the same ontological framework.

Within this sociotechnical framework, integrating values into AI is not about "embedding" ethical principles into technological artifacts. Instead, it involves recognizing how technology and society collaboratively shape and co-produce meanings and values. This perspective suggests that values are not inherently "inside" AI but are instead attributed to technology by humans through their interactions and within specific contexts. Consequently, the ethical considerations in AI development extend beyond technical issues of coding or design and require a broader approach that considers the intricate interplay between technology and society.

Following this sociotechnical approach, we can recognize that achieving those ethical considerations requires a contextualized approach to ethics in AI. One alternative is emphasizing the importance of applying a continuous hermeneutic process [12]. A hermeneutic process is not static; it evolves as new insights and understandings emerge. This iterative process means that interpretation is ongoing, with each new layer of understanding influencing and reshaping prior interpretations. This includes historical, cultural, and social contexts, recognizing that these factors influence meanings and cannot be fully understood in isolation. Furthermore, in a hermeneutic process, the interpreter must be aware of their own biases, preconceptions, and assumptions, thus examining and, where necessary, challenging these biases to allow for a more authentic and profound understanding.

Hence, by viewing AI as a sociotechnical system, ethical unveiling opens up a new dimension of convergence, where AI and ethics are not seen as separate domains but as interconnected components of a broader societal framework. In this view, AI technologies are not just tools but part of complex networks of human relations, social norms, and organizational practices.

Ethical unveiling, as presented by Arriagada-Bruneau [12], engages with a philosophical stance, Heidegger's concept of "Gestell" or technological framing, which describes how modern technology shapes our relationship with the world, reducing everything to a resource available for exploitation. According to Heidegger, this framing limits our understanding of technology's true essence, confining it to its instrumental value. In the context of AI, this technological framing can manifest as "techno-chauvinism" or "technological solutionism," where technology is seen as the ultimate solution to complex social problems, often overlooking the underlying ethical, political, and social dimensions. An ethical unveiling seeks to emancipate AI from this technological framing, encouraging a relationship with technology that goes beyond its utilitarian function. This approach challenges the notion that AI's value lies solely in its capacity to optimize processes and make decisions. Instead, it advocates for AI as a technology that can enrich our understanding of the world and our place within it.

The practical implications of ethical unveiling are profound. It requires integrating ethical considerations from the very inception of AI development, ensuring that ethics are not an afterthought but a fundamental component of the design process. This integration demands ongoing dialogue among developers, ethicists, policymakers, and the broader public, fostering a shared understanding of AI's ethical challenges and opportunities. Furthermore, ethical unveiling calls for a dynamic and adaptive approach to AI Ethics, where ethical guidelines evolve in response to new information, perspectives, and societal changes. This approach contrasts with the static application of ethical principles, emphasizing the need for continuous ethical reflection and revision throughout the lifecycle of AI systems. By embracing a hermeneutic process, AI ethics can move from being reactive and situation-specific to proactive and integral to technological innovation.

To effectively apply the concept of ethical unveiling as a hermeneutic process, one should adopt a dynamic and contextually grounded approach that integrates ethics as an intrinsic component from the beginning of AI projects. This process begins with a comprehensive contextual analysis, acknowledging the historical, cultural, and social dimensions that shape the ethical landscape in which the AI system will operate and any assumptions or limitations recognized by the developing team. Ethical unveiling should challenge the traditional technocentric paradigm in which we use more technology to fix technology and, instead, embed ethical considerations into the design phase, ensuring that the AI's purpose and operation align with societal values rather than merely technical objectives. This integration demands continuous hermeneutic reflection (profound and iterative) throughout the AI's lifecycle and its surrounding ecosystem, wherein developers, designers, users, and other stakeholders engage in iterative and continuous processes of interpretation and reinterpretation, actively questioning and refining their biases and assumptions as the project evolves. Furthermore, this engagement amongst stakeholders is crucial to ensure that diverse perspectives inform the ethical framework. This approach necessitates the development of dynamic ethical guidelines that remain adaptable to new insights and societal changes. Additionally, continuous monitoring and feedback mechanisms must be established to assess and realign the AI's performance with its ethical commitments. Through these measures, the ethical unveiling as a hermeneutic process can be applied to existing methodologies and practices within the AI field, ensuring that AI development is guided by a robust and contextually aware ethical foundation.

12.3 A preface into the future of AI

As AI continues to advance rapidly and is deployed across a wider range of fields, we outline our expectations for future developments in ethics in AI and information technology beyond the needed advancements in sustainability and reproducibility discussed in Chapter 3.

We believe that AI should be regarded as another technological revolution within the long continuum of human innovation. Like previous major technological transformations, AI presents opportunities and challenges for our economies and cultures. History shows us that technological challenges societies and induces profound transformative processes. Transformation involves changes in education [266], in the way we relate to our work, in how we interact with others, and, indeed, in countless aspects of our daily lives that are permeable to disruptive technologies. The essence of changes lies in the adaptability of institutions at the macro level, but above all, in the adaptability of individuals.

This book shows that we are experiencing a profound technological change driven by AI. Although intelligent systems have been part of our lives for years, encompassing a vast spectrum of human activities, what has characterized this interaction has been the unawareness of AI's presence. In our interaction with information systems, AI has helped us find movies or book flights. However, it has also operated in the backend of many services we use daily, such as medical image processing, fraud detection in banking, and cybersecurity threat detection. Nevertheless, this presence has been largely invisible. We have coexisted with AI without being fully aware of our interaction with it.

The technological change we are experiencing leads us to a new scenario. The presence of AI is now more acknowledged. We now know that it is an AI what makes Alexa interact with us. We now understand that when using ChatGPT, it is AI with which we are engaging. We have become aware of our interaction with AI, which is due to AI moving from the back-end of information systems to the front end, interacting directly with us in conversational systems, or assisting us in daily tasks. Its influence is now evident, and therefore much deeper.

The study of the interaction between AI and humans will likely be one of the most rapidly developing fields in the coming years. Moreover, the ability of AI to imitate humans is probably the most astonishing path that the future holds for us. With multi-agent systems powered by LLMs, we can simulate human interactions. The possibility of anticipating, through AI, what to expect from human interactions can turn fascinating. The technology at our disposal allows us to simulate meetings. Simulation enables control, and therefore, in these scenarios, it becomes more feasible to foresee the critical factors that influence the outcome of human interactions. We will increasingly hear voices advocating for replacing policymakers with intelligent agents, and decision-makers with AIs, all to reduce the pernicious effects inherent in human interaction, such as influence peddling and, ultimately, corruption. Can AI help us build a more trustworthy and transparent democracy? This is a debate that will undoubtedly emerge in the coming years.

A significant concern arises regarding the slow pace of institutional adaptation to technological change [17]. These are two processes that operate at very different dynamics. While technological development undergoes a fast change process, states grapple with liberal democracies characterized by a widespread institutional weakness and growing distrust in institutions. A key issue is how robust, credible, and reliable the application of regulatory frameworks can be in a scenario of progressive institutional weakening.

The slow pace of institutional adaptation presents complex scenarios [62]. It seems that the speed at which individuals adopt technology far outpaces the way institutions adapt their regulations to ensure the safe use of AI. A robust debate surrounding the role of multilateralism and the need to address a coordinated space for regulatory actions at a global level is crucial. It remains unclear what the outcome of this exercise will be, with initiatives coming from diverse institutional frameworks involving entities ranging from the UN [142] to the OECD [65] or UNESCO [295]. The only certainty is that these regulatory frameworks will require tremendous effort to keep them updated.

Other efforts will be necessary as well. We expect significant progress in building sociotechnical infrastructures designed to address some of the most pressing ethical challenges. For AI developers, it appears likely that emerging regulations and societal expectations will imply the development of new standards, including ethical codes and certifications. It will also lead to new practices that will require improved training to address ethical issues and, crucially, the formation of interdisciplinary teams to guide AI development. Moreover, as regulations increasingly mandate transparency and explainability, new types of professionals and organizations—potentially loosely connected—will likely emerge to represent citizens' needs and advocate for the types of AI that can (or cannot) be deployed and the degree of transparency and accountability required in each sector of our societies. To address the challenges related to employment, we anticipate a rise in evidence-based research examining the effects of AI on the job market. Although AI's impact is projected to be significant, we believe that transitioning from predictive to observational studies will provide a clearer understanding of the intricate aspects of work that cannot be entirely automated or replaced by AI. As a result, we expect a shift in focus from designing AI to replacing human roles to creating more effective and rewarding human-AI configurations of work.

Thus, the challenges we face in aligning AI with human values are vast but not insurmountable. The transformation AI brings to society is an opportunity to reimagine how technology can serve the common good, fostering a more transparent, equitable, and just future. The true potential of AI lies not only in its capacity to automate tasks or generate knowledge but also in its ability to elevate human agency and decision-making. By ensuring that ethical principles are embedded in every facet of the AI ecosystem, we can navigate this technological revolution in a way that enhances, rather than diminishes, our humanity. It is our collective responsibility to shape AI into a force for positive, transformative change.

Integrating ethics into AI requires a deeply interdisciplinary evolution in the coming years. In this book, we assert that this integration goes far beyond merely adding ethical tools to the AI toolkit. It calls for a fundamental and decisive shift in how we conceive and develop AI. Recognizing AI systems as sociotechnical constructs, the integration of ethics demands a holistic approach that addresses both the technological and societal dimensions of AI. We hope that this book has helped convey the urgency of this transformation, inspiring us all to become active agents of change in shaping a more ethical and responsible future for AI.

References

[1] Access Now. 2024. Regulatory mapping on Artificial Intelligence in Latin America: Regional AI public policy report. Technical Report, Access Now.

[2] Amina Adadi and Mohammed Berrada. 2018. Peeking inside the black-box: A survey on explainable Artificial Intelligence (XAI). IEEE Access, 6: 52138–52160.

[3] Angus Addlesee. 2024. Grounding llms to in-prompt instructions: Reducing hallucinations caused by static pre-training knowledge. 1–7.

[4] David Adkins, Bilal Alsallakh, Adeel Cheema, Narine Kokhlikyan, Emily McReynolds, et al. 2022. Method cards for prescriptive machine-learning transparency. *In*: 2022 IEEE/ACM 1st International Conference on AI Engineering – Software Engineering for AI (CAIN). 90–100.

[5] United Nations AI advisory board. 2023. Interim report: Governing AI for humanity. https://www.un.org/en/ai-advisory-body.

[6] Amazon. 2024. Responsible AI. `https://aws.amazon.com/es/machine-learning/responsible-ai/`. Accessed: 2024-05-01.

[7] Lindsey Andersen. 2020. Human rights in the age of Artificial Intelligence. International Review of the Red Cross, 102(913): 345–376.

[8] McKane Andrus and Sarah Villeneuve. 2022. Demographic-reliant algorithmic fairness: Characterizing the risks of demographic data collection in the pursuit of fairness. *In*: Proceeding of the 2022 ACM Conference on Fairness, Accountability, and Transparency, FAccT '22. 1709–1721. New York, NY, USA. Association for Computing Machinery.

[9] Julia Angwin, Jeff Larson, Surya Mattu, and Lauren Kirchner. 2016. Machine bias. ProPublica.

[10] Anjum and Rahul Katarya. 2022. Automated news summarization using transformers. *In*: Sustainable Advanced Computing. Springer.

[11] Gabriela Arriagada-Bruneau. 2024. Los sesgos del algoritmo; la importancia de diseñar una Inteligencia Artificial ética e inclusiva. La Pollera.

[12] Gabriela Arriagada-Bruneau. 2024. Una mirada crítica a la ética de la ia: de preocupaciones emergentes y principios orientadores a un desvelar ético. Resonancias. Revista De Filosofía. 17: 101–120.

[13] Alejandro Barredo Arrieta, Natalia Díaz-Rodríguez, Javier Del Ser, Adrien Bennetot, Siham Tabik, et al. 2020. Explainable Artificial Intelligence (XAI): Concepts, taxonomies, opportunities and challenges toward responsible AI. Information Fusion. 58: 82–115.

[14] Samantha Artiga and Kendal Orgera. 2016. Key facts on health and health care by race and ethnicity. Kaiser Family Foundation. 7.

[15] UN General Assembly. 2024. Seizing the opportunities of safe, secure and trustworthy Artificial Intelligence systems for sustainable development: resolution/adopted by the general assembly.

[16] Ricardo Baeza-Yates. 2018. Bias on the web. Communications of the ACM. 61(6): 54–61.

[17] Ricardo Baeza-Yates and Usama M. Fayyad. 2024. Responsible AI: an urgent mandate. IEEE Intelligent Systems. 39(1): 12–17.

[18] Zinzi D. Bailey, Nancy Krieger, Madina Agénor, Jasmine Graves, Natalia Linos, et al. 2017. Structural racism and health inequities in the USA: evidence and interventions. The lancet. 389(10077): 1453–1463.

[19] Aparna Balagopalan, Haoran Zhang, Kimia Hamidieh, Thomas Hartvigsen, Frank Rudzicz, et al. 2022. The road to explainability is paved with bias: Measuring the fairness of explanations. *In*: Proceeding of the 2022 ACM Conference on Fairness, Accountability, and Transparency, FAccT '22. 1194–1206. New York, NY, USA. Association for Computing Machinery.

[20] Gagan Bansal, Tongshuang Wu, Joyce Zhou, Raymond Fok, Besmira Nushi, et al. 2021. Does the whole exceed its parts? The effect of AI explanations on complementary team performance. *In*: Proceeding of the 2021 CHI Conference on Human Factors in Computing Systems. 1–16.

[21] Solon Barocas and Danah Boyd. 2017. Engaging the ethics of data science in practice. Communications of the ACM. 60(11): 23–25.

[22] Solon Barocas and Andrew D. Selbst. 2016. Big data's disparate impact. California Law Review. 104: 671.

[23] Christoph Bartneck, Christoph Lütge, Alan Wagner, and Sean Welsh. 2021. An introduction to Ethics in Robotics and AI. Springer Nature.

[24] Rufus Barker Bausell. 2021. The Problem with Science: The Reproducibility Crisis and What to Do About It. Oxford University Press.

[25] Gordon Baxter and Ian Sommerville. 2011. Socio-technical systems: From design methods to systems engineering. Interacting with Computers. 23(1): 4–17.

[26] BeautyAI. 2016. The first international beauty contest. https://beauty.ai/.

[27] Emily M. Bender, Timnit Gebru, Angelina McMillan-Major, and Shmargaret Shmitchell. 2021. On the dangers of stochastic parrots: Can language models be too big? *In*: Proceedings of the 2021 ACM Conference on Fairness, Accountability, and Transparency. 610–623.

[28] Ruha Benjamin. 2019. Race after Technology: Abolitionist Tools for the New Jim Code. John Wiley & Sons.

[29] Christopher Bennet. 2010. What is this thing called Ethics? Routledge, 1st edition.

[30] Joseph R. Biden. 2023. Executive order on the safe, secure, and trustworthy development and use of Artificial Intelligence. The White House.

[31] Elettra Bietti. 2020. From ethics washing to ethics bashing: A view on tech ethics from within moral philosophy. *In*: Proceedings of the 2020 Conference on Fairness, Accountability, and Transparency, FAT* '20. 210. New York, NY, USA. Association for Computing Machinery.

[32] Celeste Biever. 2023. ChatGPT broke the turing test—the race is on for new ways to assess AI. Nature. 619(7971): 686–689.

[33] Wiebe E. Bijker. 1995. Of Bicycles, Bakelites, and Bulbs: Toward a Theory of Sociotechnical Change. MIT Press, Cambridge, Massachusetts.

[34] Paula Boddington. 2023. AI Ethics. Artificial Intelligence: Foundations, Theory, and Algorithms. Springer Singapore, 1st edition.

[35] Tolga Bolukbasi, Kai-Wei Chang, James Zou, Venkatesh Saligrama, and Adam Kalai. 2016. Man is to computer programmer as woman is to homemaker? debiasing word embeddings. *In*: Proceeding of the 30th International Conference on Neural Information Processing Systems, NIPS'16. 4356–4364. Red Hook, NY, USA. Curran Associates Inc.

[36] Tolga Bolukbasi, Kai-Wei Chang, James Y. Zou, Venkatesh Saligrama, and Adam T. Kalai. 2016. Man is to computer programmer as woman is to homemaker? debiasing word embeddings. Advances in Neural Information Processing Systems. 29.

[37] Jason Borenstein and Ayanna Howard. 2020. Emerging challenges in AI and the need for AI ethics education. AI and Ethics. 1: 1–5.

[38] Leonie N. Bossert and Thilo Hagendorff. 2023. The ethics of sustainable AI: Why animals (should) matter for a sustainable use of AI. Sustainable Development. 31(5): 3459–3467.

[39] Mark Bovens. 2007. Analysing and assessing accountability: A conceptual framework. European Law Journal. 13: 447–468.

[40] Robert Brauneis and Ellen P. Goodman. 2018. Algorithmic transparency for the smart city. Yale Journal of Law & Technology. 20: 103.

[41] Andrew Brock, Jeff Donahue, and Karen Simonyan. 2019. Large scale gan training for high fidelity natural image synthesis. *In*: International Conference on Learning Representations.

[42] Meredith Broussard. 2018. Artificial unintelligence: How computers misunderstand the world. MIT Press, Cambridge, MA.

[43] Meredith Broussard. 2023. More than a Glitch: Confronting Race, Gender, and Ability Bias in Tech. MIT Press, Cambridge, MA.

[44] Tom B. Brown, Benjamin Mann, Nick Ryder, Melanie Subbiah, Jared Kaplan, et al. 2020.Language models are few-shot learners. *In*: Advances in Neural Information Processing Systems 33: Annual Conference on Neural Information Processing Systems 2020, NeurIPS 2020, December 6-12, 2020, virtual.

[45] Sébastien Bubeck, Varun Chandrasekaran, Ronen Eldan, Johannes Gehrke, Eric Horvitz, et al. 2023. Sparks of Artificial General Intelligence: Early experiments with GPT-4. https://arxiv.org/abs/2303.12712.

[46] Miriam C. Buiten. 2019. Towards intelligent regulation of Artificial Intelligence. European Journal of Risk Regulation. 10(1): 41–59.

[47] Joy Buolamwini and Timnit Gebru. 2018. Gender shades: Intersectional accuracy disparities in commercial gender classification. *In*: Conference on Fairness, Accountability, and Transparency. 77–91.

[48] Laura Cabello, Anna Katrine Jørgensen, and Anders Søgaard. 2023. On the independence of association bias and empirical fairness in language models. *In*: Proceeding of the 2023 ACM Conference on Fairness, Accountability, and Transparency, FAccT '23. 370–378. New York, NY, USA. Association for Computing Machinery.

[49] Simon Caton and Christian Haas. 2024. Fairness in machine learning: A survey. ACM Computing Surveys. 56(7).

[50] José-Antonio Cervantes, Sonia Calaza López, Luis-Felipe Rodríguez, Salvador Cervantes, Francisco Cervantes, et al. 2020. Artificial moral agents: a survey of the current status. Science and Engineering Ethics. 26(2): 501–532.

[51] Cecilia Ka Yuk Chan and Wenjie Hu. 2023. Students' voices on generative AI: Perceptions, benefits, and challenges in higher education. International Journal of Educational Technology in Higher Education. 20(1):43.

[52] Danton S. Char, Michael D. Abràmoff, and Chris Feudtner. 2020. Identifying ethical considerations for machine learning healthcare applications. The American Journal of Bioethics. 20(11): 7–17.

[53] Quanze Chen, Tobias Schnabel, Besmira Nushi, and Saleema Amershi. 2022. HINT: Integration testing for AI-based features with humans in the loop. *In*: IUI '22: 27th International Conference on Intelligent User Interfaces. 1–17. Helsinki, Finland, 2022. ACM.

[54] Robert Cherinka and J. Prezzama. 2023. The role of generative Artificial Intelligence as research assistant: Opportunities and challenges. 2023: 89–94.

[55] Neo Chilwane. 2021. Ethical Considerations for Employees Disrupted by Job Automation Technology. PhD thesis, Available from ProQuest Dissertations & Theses Global.

[56] Amit K. Chopra and Munindar P. Singh. 2018. Sociotechnical systems and ethics in the large. *In*: Proceedings of the 2018 AAAI/ACM Conference on AI, Ethics, and Society, AIES '18.48–53, New York, NY, USA. Association for Computing Machinery.

[57] Alexandra Chouldechova. 2017. Fair prediction with disparate impact: A study of bias in recidivism prediction instruments. Big Data. 5(2): 153–163.

[58] Carlos A. Cifuentes, Maria Jose Pinto Bernal, Nathalia Céspedes, and Marcela Múnera. 2020. Social robots in therapy and care. Current Robotics Reports. 1: 59–74.

[59] Danielle Keats Citron. 2022. The fight for privacy: Protecting dignity, identity, and love in the digital age. WW Norton & Company.

[60] Mark Coeckelbergh. 2010. Robot rights? Towards a social-relational justification of moral consideration. Ethics and Information Technology. 12: 209–221.

[61] European Comission. 2021. Laying down harmonised rules on Artificial Intelligence (Artificial Intelligence Act) and amending certain union legislative acts. Accessed: 2024-05-01.

[62] Marios Constantinides, Mohammad Tahaei, Daniele Quercia, Simone Stumpf, Michael Madaio, et al. 2024. Implications of regulations on the use of AI and generative AI for human-centered responsible Artificial Intelligence. *In*: Florian 'Floyd' Mueller, Penny Kyburz, Julie R. Williamson, and Corina Sas, editors, Extended Abstracts of the CHI Conference on Human Factors in Computing Systems, CHI EA 2024, Honolulu, HI, USA, May 11-16, 2024. 582: 1–582:4. ACM.

[63] Sam Corbett-Davies, Emma Pierson, Avi Feller, Sharad Goel, and Aziz Huq. 2017. Algorithmic decision making and the cost of fairness. *In*: Proceedings of the 23rd ACM SIGKDD International Conference on

Knowledge Discovery and Data Mining, KDD '17. 797–806, New York, NY, USA. Association for Computing Machinery.

[64] Sasha Costanza-Chock. 2020. Design justice: Community-led practices to build the worlds we need. The MIT Press.

[65] OECD Council on Artificial Intelligence. 2019. OECD AI principles. https://legalinstruments.oecd.org/en/instruments/OECD-LEGAL-0449.

[66] OECD Council on Artificial Intelligence. 2024. Policies, data and analysis for trustworthy Artificial Intelligence. https://oecd.ai/en/.

[67] Kate Crawford. 2021. The atlas of AI: Power, politics, and the planetary costs of Artificial Intelligence. Yale University Press.

[68] Caroline Criado-Perez. 2019. Invisible women: Data bias in a world designed for men. Abrams

[69] Russell S. Cropanzano and Maureen L. Ambrose, editors. 2015. The Oxford Handbook of Justice in the Workplace. Oxford University Press, New York.

[70] Brian d'Alessandro, Cathy O'Neil, and Tom LaGatta. 2017. Conscientious classification: A data scientist's guide to discrimination-aware classification. Big data. 5(2): 120–134.

[71] Craig M. Dalton, Linnet Taylor, and Jim Thatcher. 2016. Critical data studies: A dialog on data and space. Big Data and Society. 3.

[72] Malene F. Damholdt, Marco Nørskov, Ryuji Yamazaki, Raul Hakli, Catharina Vesterager Hansen, et al. 2015. Attitudinal change in elderly citizens toward social robots: the role of personality traits and beliefs about robot functionality. Frontiers in Psychology. 6: 1701.

[73] John Danaher. 2019. Building Better Sex Robots: Lessons from Feminist Pornography. 133–147. Springer International Publishing, Cham.

[74] Data Science for Public Policy. 2024. Aequitas: Fairness tool. Accessed: 2024-08-27.

[75] Clayton Davis, Onur Varol, Emilio Ferrara, Alessandro Flammini, and Filippo Menczer. 2016. Botornot: A system to evaluate social bots. *In*: Proceedings of the 25th International Conference Companion on World Wide Web.

[76] Matthew C. Davis, Rose Challenger, Dharshana N.W. Jayewardene, and Chris W. Clegg. 2014. Advancing socio-technical systems thinking: A call for bravery. Applied Ergonomics. 45(2, Part A): 171–180. Advances in Socio-Technical Systems Understanding and Design: A Festschrift in Honour of K.D. Eason.

[77] Jacob Devlin, Ming-Wei Chang, Kenton Lee, and Kristina Toutanova. 2019. Bert: Pre-training of deep bidirectional transformers for language understanding. *In*: Proceeding of the North American

Chapter of the Association for Computational Linguistics: Human Language Technologies, Volume 1 (Long and Short Papers). 4171–4186. Association for Computational Linguistics.

[78] Sanchari Dhar and Lior Shamir. 2021. Evaluation of the benchmark datasets for testing the efficacy of deep convolutional neural networks. Visual Informatics. 5(3): 92–101.

[79] Prafulla Dhariwal and Alex Nichol. 2021. Diffusion models beat GANs on image synthesis. *In*: Advances in Neural Information Processing Systems (NeurIPS). 34: 8780–8794.

[80] William Dieterich, Christina Mendoza, and Tim Brennan. 2016. COMPAS risk scales: Demonstrating accuracy equity and predictive parity. Technical Report, Northpointe Inc. Research Department. Performance of the COMPAS Risk Scales in Broward County.

[81] Catherine D'ignazio and Lauren F. Klein. 2023. Data feminism. MIT Press.

[82] European disability Forum. 2017. Universal design. https://rm.coe.int/presentation-rodolfo-cattani/168076474c.

[83] Mateusz Dolata, Stefan Feuerriegel, and Gerhard Schwabe. 2022. A sociotechnical view of algorithmic fairness. Information Systems Journal. 32(4): 754–818.

[84] Alexey Dosovitskiy, Lucas Beyer, Alexander Kolesnikov, Dirk Weissenborn, Xiaohua Zhai, et al. 2021. An image is worth 16×16 words: Transformers for image recognition at scale. *In*: 9th International Conference on Learning Representations, ICLR 2021, Virtual Event, Austria, May 3-7, 2021. OpenReview.net.

[85] Claude Draude, Gabriele Klumbyte, Philipp Lücking, and Pat Treusch. 2020. Situated algorithms: a sociotechnical systemic approach to bias. Online Information Review. 44(2): 325–342.

[86] Shitong Duan, Xiaoyuan Yi, Peng Zhang, Tun Lu, Xing Xie, et al. 2024. Denevil: towards deciphering and navigating the ethical values of large language models via instruction learning. *In*: International Conference on Learning Representations (ICLR).

[87] Paul Dumouchel and Luisa Damiano. 2017. Living with Robots. Harvard University Press, Cambridge, MA and London, England.

[88] Paul Aldrin Pineda Dungca. 2023. The incorporation of Large Language Models (LLMs) in the field of education: Ethical possibilities, threats, and opportunities.

[89] Cynthia Dwork, Moritz Hardt, Toniann Pitassi, Omer Reingold, and Richard Zemel. 2012. Fairness through awareness. *In*: Proceeding of the 3rd Innovations in Theoretical Computer Science Conference, ITCS '12. 214–226. New York, NY, USA. Association for Computing Machinery.

[90] Cynthia Dwork and Aaron Roth. 2014. The algorithmic foundations of differential privacy. Foundations and Trends in Theoretical Computer Science. 9(3–4): 211–407.

[91] Upol Ehsan and Mark O. Riedl. 2020. Human-centered explainable AI: Towards a reflective sociotechnical approach. *In*: HCI International 2020-Late Breaking Papers: Multimodality and Intelligence: 22nd HCI International Conference, HCII 2020, Copenhagen, Denmark, July 19–24, 2020, Proceedings 22. 449–466. Springer.

[92] Upol Ehsan and Mark O. Riedl. 2024. Explainability pitfalls: Beyond dark patterns in explainable AI. Patterns. 5(6).

[93] Upol Ehsan, Philipp Wintersberger, Q Vera Liao, Elizabeth Anne Watkins, Carina Manger, et al. 2022. Human-centered explainable AI (HCXAI): beyond opening the black-box of AI. *In*: CHI Conference on Human Factors in Computing Systems Extended Abstracts. 1–7.

[94] Fred Emery. 1980. Designing socio-technical systems for 'greenfield' sites. Journal of Occupational Behaviour. 1(1): 19–27.

[95] Fred E. Emery and Eric L. Trist. 1960. Socio-technical systems. Management Science, Models and Techniques. 2: 83–97.

[96] Patrick Esser, Sumith Kulal, Andreas Blattmann, Rahim Entezari, Jonas Müller, et al. 2024. Scaling rectified flow transformers for high-resolution image synthesis. *In*: International Conference on Machine Learning (ICML). 503.

[97] Virginia Eubanks. 2018. Automating inequality: How high-tech tools profile, police, and punish the poor. St. Martin's Press.

[98] European Parliament. 2023. Artificial Intelligence Act. Accessed: 2024-05-01.

[99] Michael Feldman, Sorelle A. Friedler, John Moeller, Carlos Scheidegger, and Suresh Venkatasubramanian. 2015. Certifying and removing disparate impact. *In*: Proceedings of the 21th ACM SIGKDD International Conference on Knowledge Discovery and Data Mining, KDD '15. 259–268. New York, NY, USA. Association for Computing Machinery.

[100] Emilio Ferrara. 2024. GenAI against humanity: nefarious applications of generative Artificial Intelligence and large language models. Journal of Computational Social Science.

[101] Jessica Fjeld, Nele Achten, Hannah Hilligoss, Adam Nagy, and Madhulika Srikumar. 2020. Principled Artificial Intelligence: Mapping consensus in ethical and rights-based approaches to principles for AI. SSRN Electronic Journal.

[102] Luciano Floridi, Josh Cowls, Monica Beltrametti, Raja Chatila, Patrice Chazerand, et al. 2018. AI4People—an ethical framework for a good AI

society: opportunities, risks, principles, and recommendations. Minds and Machines. 28(4): 689–707.

[103] Luciano Floridi and Jeff Sanders. 2004. On the morality of artificial agents. Minds and Machines. 14: 349–379.

[104] Elizabeth Ford, Richard Milne, and Keegan Curlewis. 2023. Ethical issues when using digital biomarkers and artificial intelligence for the early detection of dementia. WIREs Data Mining and Knowledge Discovery. 13(3): e1492.

[105] Eduard Fosch-Villaronga and Jordi Albo-Canals. 2019. Reflecting upon the legal and ethical aspects of the use and development of social robots for therapy. Paladyn, Journal of Behavioral Robotics. 10(1): 77–93.

[106] Sorelle A. Friedler, Carlos Scheidegger, and Suresh Venkatasubramanian. 2021. The (im) possibility of fairness: Different value systems require different mechanisms for fair decision making. Communications of the ACM. 64(4):136–143.

[107] Sorelle A. Friedler, Carlos Scheidegger, Suresh Venkatasubramanian, Sonam Choudhary, Evan P. Hamilton, et al. 2019. A comparative study of fairness-enhancing interventions in machine learning. *In*: Proceedings of the Conference on Fairness, Accountability, and Transparency, FAT* '19. 329–338, New York, NY, USA. Association for Computing Machinery.

[108] Iason Gabriel, Arianna Manzini, Geoff Keeling, Lisa Anne Hendricks, Verena Rieser, et al. 2024. The ethics of advanced AI assistants. https://arxiv.org/abs/2404.16244.

[109] Jai Galliott. 2015. Military Robots: Mapping the Moral Landscape. Routledge, 1st edition.

[110] Ruoyuan Gao and Chirag Shah. 2021. Addressing bias and fairness in search systems. *In*: Proceedings of the 44th International ACM SIGIR Conference on Research and Development in Information Retrieval, SIGIR '21. 2643–2646. New York, NY, USA. Association for Computing Machinery.

[111] Zhengjie Gao, Xuanzi Liu, Yuanshuai Lan, and Zheng Yang. 2024. A brief survey on safety of large language models. Journal of Computing and Information Technology. 32(1): 47–64.

[112] Patrick Gebhard, Ruth Aylett, Ryuichiro Higashinaka, Kristiina Jokinen, Hiroki Tanaka, et al. 2021. Modeling Trust and Empathy for Socially Interactive Robots. 21–60. Springer Singapore, Singapore.

[113] Timnit Gebru, Jamie Morgenstern, Briana Vecchione, Jennifer Wortman Vaughan, Hanna Wallach, et al. 2021. Datasheets for datasets. Communications of the ACM. 64(12): 86–92.

[114] Gonzalo Génova, Valentín Moreno, and M Rosario González. 2023. Machine ethics: do androids dream of being good people? Science and Engineering Ethics. 29(2): 10.

[115] Sara Gerke. 2023. "Nutrition Facts Labels" for Artificial Intelligence/Machine Learning-based medical devices-the urgent need for labeling standards. George Washington Law Review. 91: 79.

[116] Bishwamittra Ghosh, Debabrota Basu, and Kuldeep S. Meel. 2023. "How biased are your features?": Computing fairness influence functions with global sensitivity analysis. *In*: Proceedings of the 2023 ACM Conference on Fairness, Accountability, and Transparency, FAccT '23. 138–148. New York, NY, USA. Association for Computing Machinery.

[117] Aidan Gilson, Conrad W. Safranek, Thomas Huang, Vimig Socrates, Ling Chi, et al. 2023. How does ChatGPT perform on the United States medical licensing examination? the implications of large language models for medical education and knowledge assessment. JMIR Medical Education. 9.

[118] Trystan S. Goetze. 2023. Integrating ethics into computer science education: Multi-, inter-, and transdisciplinary approaches. *In*: Proceedings of the 54th ACM Technical Symposium on Computer Science Education V. 1, SIGCSE 2023. 645–651, New York, NY, USA. Association for Computing Machinery.

[119] Ian J. Goodfellow, Jean Pouget-Abadie, Mehdi Mirza, Bing Xu, David Warde-Farley, et al. 2014. Generative adversarial nets. *In*: Advances in Neural Information Processing Systems (NIPS). 27: 2672–2680.

[120] Google. 2023. Measuring the user experience on a large scale: User-centered metrics for web applications.

[121] Google. 2023. Responsible AI practices. `https://ai.google/responsibility/responsible-ai-practices/`. Accessed: 2023-05-01.

[122] GPTZero. 2024. More than an AI detector preserve what's human. `https://gptzero.me/`.

[123] Simone Grassini. 2023. Shaping the future of education: Exploring the potential and consequences of AI and ChatGPT in educational settings. Education Sciences. 13(7).

[124] Mary L. Gray and Siddharth Suri. 2019. Ghost work: How to stop Silicon Valley from building a new global underclass. Eamon Dolan Books.

[125] Zacharus Gudmunsen. 2024. The moral decision machine: A challenge for artificial moral agency based on moral deference. AI Ethics.

[126] Odd Erik Gundersen. 2021. The fundamental principles of reproducibility. Philosophical Transactions of the Royal Society A, 379(2197): 20200210.

[127] Odd Erik Gundersen and Sigbjørn Kjensmo. 2018. State of the art: Reproducibility in Artificial Intelligence. *In*: Proceeding of the AAAI Conference on Artificial Intelligence. Volume 32.

[128] David J. Gunkel. 2018. The other question: Can and should robots have rights? Ethics and Information Technology. 20(2): 87–99.

[129] Thilo Hagendorff. 2020. The ethics of AI ethics: An evaluation of guidelines. Minds and Machines. 30: 99.

[130] Bingyi Han, Sadia Nawaz, George Buchanan, and Dana McKay. 2023. Ethical and pedagogical impacts of AI in education. *In*: Ning Wang, Genaro Rebolledo-Mendez, Noboru Matsuda, Olga C. Santos, and Vania Dimitrova, editors, Artificial Intelligence in Education. 667–673, Cham. Springer Nature Switzerland.

[131] Sandra Harding. 1995. "Strong objectivity": A response to the new objectivity question. Synthese. 104: 331–349.

[132] Sandra Harding. 2019. Objectivity and diversity: Another logic of scientific research. University of Chicago Press.

[133] Sandra Harding. 2023. Science and social inequality: Feminist and postcolonial issues. University of Illinois Press.

[134] Moritz Hardt, Eric Price, and Nathan Srebro. 2016. Equality of opportunity in supervised learning. *In*: Proceedings of the 30th International Conference on Neural Information Processing Systems, NIPS'16. 3323–3331, Red Hook, NY, USA. Curran Associates Inc.

[135] Thomas Hartvigsen, Saadia Gabriel, Hamid Palangi, and Maarten Sap. 2022. Toxigen: A large-scale machine-generated dataset for adversarial and implicit hate speech detection. *In*: Proceedings of the 60th Annual Meeting of the Association for Computational Linguistics. 3309–3326.

[136] David Haussler. 1988. Quantifying inductive bias: AI learning algorithms and valiant's learning framework. Artificial Intelligence. 36(2): 177–221.

[137] Kaiming He, Xiangyu Zhang, Shaoqing Ren, and Jian Sun. 2016. Deep residual learning for image recognition. *In*: Proceedings of the IEEE Conference on Computer Vision and Pattern Recognition. 770–778.

[138] Thomas Hellström. 2013. On the moral responsibility of military robots. Ethics and Information Technology. 15: 99–107.

[139] HLEG. 2018. The assessment list for trustworthy Artificial Intelligence (altai). https://futurium.ec.europa.eu/en/european-ai-alliance/pages/welcome-altai-portal.

[140] HLEG. 2018. Ethics guidelines for trustworthy AI. https://digital-strategy.ec.europa.eu/en/library/ethics-guidelines-trustworthy-ai.

[141] Jonathan Ho, Ajay Jain, and Pieter Abbeel. 2020. Denoising diffusion probabilistic models. *In*: Advances in Neural Information Processing Systems. 33: 6840–6851.

[142] Lambert Hogenhout. 2021. A framework for ethical AI at the United Nations, unite paper (1).

[143] Lei Huang, Weijiang Yu, Weitao Ma, Weihong Zhong, Zhangyin Feng, et al. 2023. A survey on hallucination in large language models: Principles, taxonomy, challenges, and open questions. https://arxiv.org/abs/2311.05232.

[144] Eyke Hüllermeier, Thomas Fober, and Margret Mernberger. 2013. Inductive bias. *In*: Werner Dubitzky, Olaf Wolkenhauer, Kwang-Hyun Cho, and Hiroki Yokota, editors, Encyclopedia of Systems Biology. Springer, New York, NY.

[145] Benjamin Hunter, Sumeet Hindocha, and Richard W. Lee. 2022. The role of Artificial Intelligence in early cancer diagnosis. Cancers. 14(6).

[146] IBM. 2022. Everyday ethics for AI. Accessed: 2023-05-31.

[147] Andrew Iliadis and Federica Russo. 2016. Critical data studies: An introduction. Big Data and Society.

[148] Innovation, Science and Economic Development Canada. 2022. Artificial Intelligence and Data Act (AIDA) companion document. `https: //ised-isde.canada.ca/site/innovation-better-canada/en/artificial-intelligence-and-data-act-aida-companion-document`. Accessed: 2024-05-01.

[149] Innovation, Science and Economic Development Canada. 2023. Canadian guardrails for generative AI: Code of practice. https://ised-isde.canada.ca/site/ised/en/consultation-development-canadian-code-practice-generative-artificial-intelligence-systems/canadian-guardrails-generative-ai-code-practice. Accessed: 2023-05-01.

[150] Jim Isaak and Mina J. Hanna. 2018. User data privacy: Facebook, Cambridge Analytica, and privacy protection. Computer. 51(8): 56–59.

[151] Elizabeth Jackson, Tyron Goldschmidt, Dustin Crummett, and Rebecca Chan. 2021. Applied ethics: An impartial introduction. Hackett Publishing.

[152] Maia Jacobs, Melanie F. Pradier, Thomas H. McCoy Jr, Roy H. Perlis, Finale Doshi-Velez, et al. 2021. How machine-learning recommendations influence clinician treatment selections: the example of antidepressant selection. Translational Psychiatry. 11(1): 108.

[153] Prachi Jain, Ashutosh Sathe, Varun Gumma, Kabir Ahuja, and Sunayana Sitaram. 2024. Mafia: Multi-adapter fused inclusive language models this paper has content that might be offensive, or upsetting, however, this cannot be avoided owing to the nature of the work. *In*: EACL 2024 - 18th Conference of the European Chapter of the Association for Computational Linguistics, Proceedings of the Conference, 1: 627–645.

[154] Maurice Jakesch, Zana Buçinca, Saleema Amershi, and Alexandra Olteanu. 2022. How different groups prioritize ethical values for responsible AI. *In*: Proceedings of the 2022 ACM Conference on Fairness, Accountability, and Transparency, FAccT '22. 310–323. New York, NY, USA. Association for Computing Machinery.

[155] Ziwei Ji, Nayeon Lee, Rita Frieske, Tiezheng Yu, Dan Su, et al. 2023. Survey of hallucination in natural language generation. ACM Computing Surveys. 55(12).

[156] Jinglu Jiang, Surinder Kahai, and Ming Yang. 2022. Who needs explanation and when? juggling explainable AI and user epistemic uncertainty. International Journal of Human-Computer Studies. 165: 102839.

[157] Anna Jobin, Marcello Ienca, and Effy Vayena. 2019. The global landscape of AI ethics guidelines. Nature machine intelligence. 1(9): 389–399.

[158] Deborah Johnson and Mario Verdicchio. 2023. Computing ethics: Ethical AI is not about AI. Communications of the ACM. 66(2): 32–34.

[159] Deborah G. Johnson and Mario Verdicchio. 2017. Reframing AI discourse. Minds and Machines. 27: 575–590.

[160] Deborah G. Johnson and Mario Verdicchio. 2024. The sociotechnical entanglement of AI and values. AI and SOCIETY. 1–10.

[161] Przemyslaw Joniak and Akiko Aizawa. 2022. Gender biases and where to find them: Exploring gender bias in pre-trained transformer-based language models using movement pruning. *In*: GeBNLP 2022 - 4th Workshop on Gender Bias in Natural Language Processing, Proceedings of the Workshop. 67–73.

[162] Faisal Kamiran and Toon Calders. 2012. Data preprocessing techniques for classification without discrimination. Knowledge and Information Systems. 33(1): 1–33.

[163] Toshihiro Kamishima, Shotaro Akaho, Hideki Asoh, and Jun Sakuma. 2012. Fairness-aware classifier with prejudice remover regularizer. *In*: European Conference on Machine Learning and Knowledge Discovery in Databases. 35–50. Springer.

[164] Sayash Kapoor and Arvind Narayanan. 2023. Leakage and the reproducibility crisis in machine-learning-based science. Patterns. 4(9).

[165] Balaram Yadav Kasula. 2023. Ethical considerations in the adoption of Artificial Intelligence for mental health diagnosis 2023. International Journal of Creative Research in Computer Technology and Design. 5(5): 1–7.

[166] Michael Kearns and Aaron Roth. 2019. The Ethical Algorithm: The Science of Socially Aware Algorithm Design. Oxford University Press.

[167] Arif Ali Khan, Sher Badshah, Peng Liang, Muhammad Waseem, Bilal Khan, et al. 2022. Ethics of AI: A systematic literature review of principles and challenges. *In*: Proceedings of the 26th International Conference on Evaluation and Assessment in Software Engineering. 383–392.

[168] Faiz Ahmad Khan, Arman Majidulla, Gamuchirai Tavaziva, Ahsana Nazish, Syed Kumail Abidi, et al. 2020. Chest x-ray analysis with deep learning-based software as a triage test for pulmonary tuberculosis: a prospective study of diagnostic accuracy for culture-confirmed disease. The Lancet Digital Health. 2(11): e573–e581.

[169] Niki Kilbertus, Mateo Rojas-Carulla, Giambattista Parascandolo, Moritz Hardt, Dominik Janzing, et al. 2017. Avoiding discrimination through causal reasoning. *In*: Proceeding of the 31st International Conference on Neural Information Processing Systems, NIPS'17. 656–666. Red Hook, NY, USA. Curran Associates Inc.

[170] Diederik P. Kingma and Max Welling. 2014. Auto-encoding variational bayes. *In*: 2nd International Conference on Learning Representations (ICLR).

[171] Lauren Klein and Catherine D'Ignazio. 2024. Data feminism for AI. *In*: The 2024 ACM Conference on Fairness, Accountability, and Transparency. 100–112.

[172] Jon Kleinberg, Sendhil Mullainathan, and Manish Raghavan. 2016. Inherent trade-offs in the fair determination of risk scores. arXiv preprint arXiv: 1609.05807.

[173] Blanka Klimova, Marcel Pikhart, and Jaroslav Kacetl. 2023. Ethical issues of the use of AI-driven mobile apps for education. Frontiers in Public Health. 10.

[174] Alina Köchling, Shirin Riazy, Marius Claus Wehner, and Katharina Simbeck. 2021. Highly accurate, but still discriminatory: A fairness evaluation of algorithmic video analysis in the recruitment context. Business & Information Systems Engineering. 63: 39–54.

[175] Nima Kordzadeh and Maryam Ghasemaghaei. 2022. Algorithmic bias: review, synthesis, and future research directions. European Journal of Information Systems. 31(3): 388–409.

[176] Tanja Kubes. 2019. New materialist perspectives on sex robots. A feminist dystopia/utopia? Social Sciences. 8(8).

[177] Matt Kusner, Joshua Loftus, Chris Russell, and Ricardo Silva. 2017. Counterfactual fairness. NIPS'17. 4069–4079. Red Hook, NY, USA. Curran Associates Inc.

[178] Thomas LaCroix. 2022. Moral dilemmas for moral machines. AI Ethics. 2: 737–746.

[179] Jeff Larson, Surya Mattu, Lauren Kirchner, and Julia Angwin. 2016. How we analyzed the COMPAS recidivism algorithm. ProPublica. Accessed: 2024-08-16.

[180] Stefan Larsson and Fredrik Heintz. 2020. Transparency in Artificial Intelligence. Internet Policy Review. 9.

[181] David Leslie. 2019. Understanding Artificial Intelligence ethics and safety. arXiv preprint arXiv:1906.05684.

[182] Hanzhou Li, John T Moon, Saptarshi Purkayastha, Leo Anthony Celi, Hari Trivedi, et al. 2023. Ethics of large language models in medicine and medical research. The Lancet Digital Health. 5(6): e333–e335.

[183] David Liu, Virginie Do, Nicolas Usunier, and Maximilian Nickel. 2023. Group fairness without demographics using social networks. *In*: Proceedings of the 2023 ACM Conference on Fairness, Accountability, and Transparency, FAccT '23. 1432–1449. New York, NY, USA. Association for Computing Machinery.

[184] Tianyu Liu, Xin Zheng, Xiaoan Ding, Baobao Chang, and Zhifang Sui. 2020. An empirical study on model-agnostic debiasing strategies for robust natural language inference. *In*: CoNLL 2020 - 24th Conference on Computational Natural Language Learning, Proceedings of the Conference. 596–608.

[185] Zhongzhou Liu, Yuan Fang, and Min Wu. 2023. Mitigating popularity bias for users and items with fairness-centric adaptive recommendation. ACM Transactions on Information Systems. 41(3).

[186] Claudia López, Gabriela Arriagada-Bruneau, and Alexandra Davidoff. 2023. ¿Cómo navegar el camino hacia la ética en IA? Revista Bits de Ciencia. (25).

[187] Alba Lozano and Carolina Blanco Fontao. 2023. Is the education system prepared for the irruption of Artificial Iintelligence? a study on the perceptions of students of primary education degree from a dual perspective: Current pupils and future teachers. Education Sciences. 13(7).

[188] Alexandra Sasha Luccioni, Sylvain Viguier, and Anne-Laure Ligozat. 2023. Estimating the carbon footprint of bloom, a 176b parameter language model. Journal of Machine Learning Research. 24(253): 1–15.

[189] Kristian Lum and William Isaac. 2016. To predict and serve? Significance, 13(5): 14–19.

[190] Scott M. Lundberg and Su-In Lee. 2017. A unified approach to interpreting model predictions. *In*: Advances in Neural Information Processing Systems. 30: 4765–4774. Curran Associates, Inc.

[191] Yougang Lyu, Piji Li, Yechang Yang, Maarten de Rijke, Pengjie Ren, et al. 2023. Feature-level debiased natural language understanding. *In*:

Proceedings of the 37th AAAI Conference on Artificial Intelligence, AAAI 2023. 37: 13353–13361.

[192] Bodhisattwa Prasad Majumder, Zexue He, and Julian McAuley. 2023. Interfair: Debiasing with natural language feedback for fair interpretable predictions. *In*: EMNLP 2023–2023 Conference on Empirical Methods in Natural Language Processing, Proceedings. 9466–9471.

[193] Charlton D. McIlwain. 2020. Black Software: The Internet and Racial Justice, from the AfroNet to Black Lives Matter. Oxford University Press, New York, NY.

[194] Stuart McLennan, Amelia Fiske, Leo Anthony Celi, Ruth Müller, Jan Harder, et al. 2020. An embedded ethics approach for AI development. Nature Machine Intelligence. 2(9): 488–490.

[195] Ninareh Mehrabi, Fred Morstatter, Nripsuta Saxena, Kristina Lerman, and Aram Galstyan. 2021. A survey on bias and fairness in machine learning. ACM Computing Surveys. 54(6).

[196] Prem Melville and Vikas Sindhwani. 2010. Recommender systems. Encyclopedia of Machine Learning. 829–838.

[197] Marcelo Mendoza, Eliana Providel, Marcelo Santos, and Sebastián Valenzuela. 2024. Detection and impact estimation of social bots in the Chilean twitter network. Scientific Reports. 14: 6525.

[198] Marcelo Mendoza, Sebastián Valenzuela, Enrique Núñez-Mussa, Fabián Padilla, Eliana Providel, et al. 2023. A study on information disorders on social networks during the chilean social outbreak and covid-19 pandemic. Applied Sciences. 13(9): 5347.

[199] Meta Platforms, Inc. 2022. Generating responsible associations. `https://ai.meta.com/responsible-ai/`. Accessed: 2024-05-01.

[200] Danaë Metaxa, Joon Sung Park, Ronald E. Robertson, Karrie Karahalios, Christo Wilson, et al. 2021. Auditing algorithms: Understanding algorithmic systems from the outside in. Foundations and Trends in Human-Computer Interaction, 14(4): 272–344.

[201] Denise Meyerson and Catriona Mackenzie. 2018. Procedural justice and the law. Philosophy Compass. 13(12): e12548.

[202] Letitia Meynell and Clarisse Paron. 2023. Applied Ethics Primer. Broadview Press.

[203] Microsoft. Responsible AI dashboard. `https://www.microsoft.com/en-us/ai/responsible-ai`. Accessed: 2023-05-01.

[204] Microsoft. Microsoft responsible AI standard, version 2. Technical report, Microsoft Corporation, June 2022. For External Release.

[205] Microsoft. Voluntary commitments by microsoft to advance responsible AI innovation, July 2023. Accessed: 2024-05-01.

[206] Tim Miller. 2019. Explanation in artificial intelligence: Insights from the social sciences. Artificial Intelligence. 267: 1–38.

[207] Tim Miller, Piers Howe, and Liz Sonenberg. 2017. Explainable AI: Beware of inmates running the asylum or: How I learnt to stop worrying and love the social and behavioural sciences. arXiv preprint arXiv:1712.00547.

[208] Margaret Mitchell, Simone Wu, Andrew Zaldivar, Parker Barnes, Lucy Vasserman, et al. 2019. Model cards for model reporting. *In*: Proceedings of the Conference On fairness, Accountability, and Transparency. 220–229.

[209] Brent Mittelstadt. 2019. Principles alone cannot guarantee ethical AI. Nature Machine Intelligence. 1: 501.

[210] James H. Moor. 2006. The nature, importance, and difficulty of machine ethics. IEEE Intelligent Systems. 21(4): 18–21.

[211] Cristian Moyano-Fernández and Jon Rueda. 2024. AI, sustainability, and environmental ethics. *In*: Ethics of Artificial Intelligence. 219–236. Springer.

[212] Marcus R. Munafò, Chris Chambers, Alexandra Collins, Laura Fortunato, and Malcolm Macleod. 2022. The reproducibility debate is an opportunity, not a crisis. BMC Research Notes. 15(1): 43.

[213] Luke Munn. 2023. The uselessness of AI ethics. AI Ethics. 3: 871.

[214] Luca Nannini. 2023. Voluminous yet vacuous? semantic capital in an age of large language models. *In*: Proceedings of the Workshop on Ethics and Trust in Human-AI Collaboration: Socio-Technical Approaches (ETHAICS 2023), co-located with 32nd International Joint Conference on Artificial Intelligence (IJCAI 2023). 115–124. CEUR-WS.org.

[215] Arvind Narayanan and Vitaly Shmatikov. 2008. Robust de-anonymization of large sparse datasets. *In*: 2008 IEEE Symposium on Security and Privacy (sp 2008). 111–125. IEEE.

[216] Lama H. Nazer, Razan Zatarah, Shai Waldrip, and Janny Xue Chen Ke. 2023. Bias in Artificial Intelligence algorithms and recommendations for mitigation. PLOS Digital Health.

[217] Philip J. Nickel. 2013. Trust and sociotechnical systems. Interacting with Computers. 25(4): 294–306.

[218] Safiya Umoja Noble. 2018. Algorithms of oppression: How search engines reinforce racism. *In*: Algorithms of Oppression. New York University Press.

[219] Northpointe. 2012. Correctional offender management profiling for alternative sanctions (COMPAS).
https://doc.wi.gov/Pages/AboutDOC/COMPAS.aspx.

[220] Caterina Novelli, Mariarosaria Taddeo, and Luciano Floridi. 2023. Accountability in Artificial Intelligence: What it is and how it works. AI and Society.

[221] N.W.2d. 2016. State of wisconsin v. loomis. Technical report, Wisconsin Supreme Court.

[222] Ziad Obermeyer, Brian Powers, Christine Vogeli, and Sendhil Mullainathan. 2019. Dissecting racial bias in an algorithm used to manage the health of populations. Science. 366(6464): 447–453.

[223] OECD. 2023. Artificial Intelligence in Science: Challenges, Opportunities and the Future of Research. Technical report, Organisation for Economic Co-operation and Development (OECD).

[224] Chinasa T. Okolo, Nicola Dell, and Aditya Vashistha. 2022. Making AI explainable in the global south: A systematic review. *In*: Proceedings of the 5th ACM SIGCAS/SIGCHI Conference on Computing and Sustainable Societies. 439–452.

[225] Alexandra Olteanu, Carlos Castillo, Fernando Diaz, and Emre Kıcıman. 2019. Social data: Biases, methodological pitfalls, and ethical boundaries. Frontiers in big data. 2: 13.

[226] International Panel on the Information Environment. 2024. IPIE submission to the UN AI advisory body. https://www.ipie.info/reports-and-publications/reports-and-publications.

[227] United Nations Conference on Trade and RM Development. 2021. Data protection and privacy legislation worldwide.

[228] Cathy O'Neil. 2017. Weapons of math destruction: How big data increases inequality and threatens democracy. Crown.

[229] OpenAI. 2023. Developing beneficial AGI safely and responsibly. Accessed: 2024-05-01.

[230] OpenAI. 2023. GPT-4 technical report. https://arxiv.org/abs/2303.08774.

[231] OpenAI. 2023. New AI classifier for indicating AI-written text. `https://openai.com/blog/new-ai-classifier-for-indicating-ai-written-text`.

[232] OpenAI. 2023. Our approach to alignment research. Accessed: 2024-05-01.

[233] Hadas Orgad, Seraphina Goldfarb-Tarrant, and Yonatan Belinkov. 2022. How gender debiasing affects internal model representations, and why it matters. *In*: NAACL 2022–2022 Conference of the North American Chapter of the Association for Computational Linguistics: Human Language Technologies, Proceedings of the Conference. 2602–2628.

[234] Long Ouyang, Jeffrey Wu, Xu Jiang, Diogo Almeida, Carroll Wainwright, et al. 2022. Training language models to follow instructions with human

feedback. *In*: Advances in Neural Information Processing Systems. 35: 27730–27744.

[235] Nicolas Papernot, Fartash Faghri, Nicholas Carlini, Ian Goodfellow, Reuben Feinman, et al. 2018. Technical report on the cleverhans v2.1.0 adversarial examples library. arXiv preprint arXiv:1610.00768.

[236] European Parliament. 2016. General data protection regulation (GDPR). https://gdpr-info.eu/.

[237] William Peebles and Saining Xie. 2023. Scalable diffusion models with transformers. *In*: Proceedings of the IEEE/CVF International Conference on Computer Vision (ICCV). 4195–4205.

[238] Andi Peng, Besmira Nushi, Emre Kiciman, and Kori Inkpen. 2022. Investigations of performance and bias in human-ai teamwork in hiring. Proceedings of the AAAI Conference on Artificial Intelligence.

[239] Sundar Pichai. 2018. AI at google: our principles. The Keyword. 7(2018): 1–3.

[240] Trevor J. Pinch and Wiebe E. Bijker. 1984. The social construction of facts and artefacts: or how the sociology of science and the sociology of technology might benefit each other. Social Studies of Science. 14(3): 399–441.

[241] Alec Radford, Jong Wook Kim, Chris Hallacy, Aditya Ramesh, Gabriel Goh, et al. 2021. Learning transferable visual models from natural language supervision. *In*: Proceedings of the 34th International Conference on Neural Information Processing Systems (NeurIPS). 8748–8763. Curran Associates Inc.

[242] Alec Radford, Karthik Narasimhan, Tim Salimans, and Ilya Sutskever. 2018. Improving language understanding by generative pre-training.

[243] Oliver Radley-Gardner, Hugh Beale, and Reinhard Zimmermann, editors. 2016. Fundamental Texts on European Private Law. Hart Publishing.

[244] Colin Raffel, Noam Shazeer, Adam Roberts, Katherine Lee, and Sharan Narang. 2020. Exploring the limits of transfer learning with a unified text-to-text transformer. Journal of Machine Learning Research. 21(140): 1–67.

[245] Krithika Ramesh, Arnav Chavan, Shrey Pandit, and Sunayana Sitaram. 2023. A comparative study on the impact of model compression techniques on fairness in language models. *In*: Anna Rogers, Jordan Boyd-Graber, and Naoaki Okazaki, editors, Proceedings of the 61st Annual Meeting of the Association for Computational Linguistics (Volume 1: Long Papers). 15762–15782, Toronto, Canada. Association for Computational Linguistics.

[246] John Rawls. 1971. A Theory of Justice. Harvard University Press, Cambridge, MA.

[247] Michael Reddy. 1979. The conduit metaphor: A case of frame conflict in our language about language. *In*: A. Ortony, editor, Metaphor and Thought. 284–324. Cambridge University Press.

[248] Google Research. 2023. Happiness tracking surveys: Large-scale in-product measurement of user attitudes and experiences. Google Research Website. Accessed: 2023-05-01.

[249] IBM Research. AI fairness 360, 2024. Accessed: 2024-08-27.

[250] Marco Tulio Ribeiro, Sameer Singh, and Carlos Guestrin. 2016. "Why should I trust you?" Explaining the predictions of any classifier. *In*: Proceedings of the 22nd ACM SIGKDD International Conference on Knowledge Discovery and Data Mining. 1135–1144.

[251] Marco Tulio Ribeiro, Sameer Singh, and Carlos Guestrin. 2016. "Why should I trust you?": Explaining the predictions of any classifier. *In*: Proceedings of the 22nd ACM SIGKDD International Conference on Knowledge Discovery and Data Mining, KDD '16. 1135–1144. New York, NY, USA. Association for Computing Machinery.

[252] Mireia Ribera and Agata Lapedriza. 2019. Can we do better explanations? a proposal of user-centered explainable AI. CEUR Workshop Proceedings.

[253] Kantwon Rogers and Ayanna Howard. 2023. Tempering transparency in human-robot interaction. 2023 IEEE International Symposium on Ethics in Engineering, Science, and Technology (ETHICS). 01–02.

[254] Robin Rombach, Andreas Blattmann, Dominik Lorenz, Patrick Esser, and Björn Ommer. 2022. High-resolution image synthesis with latent diffusion models. *In*: Proceedings of the IEEE/CVF Conference on Computer Vision and Pattern Recognition (CVPR). 10684–10695.

[255] Felipe Romero. 2019. Philosophy of science and the replicability crisis. Philosophy Compass. 14(11): e12633.

[256] Yao Rong, Tobias Leemann, Thai-Trang Nguyen, Lisa Fiedler, Peizhu Qian, et al. 2023. Towards human-centered explainable AI: A survey of user studies for model explanations. IEEE Transactions on Pattern Analysis and Machine Intelligence.

[257] Olaf Ronneberger, Philipp Fischer, and Thomas Brox. 2015. U-net: Convolutional networks for biomedical image segmentation. *In*: Nassir Navab, Joachim Hornegger, William M. Wells, and Alejandro F. Frangi, editors, Medical Image Computing and Computer-Assisted Intervention – MICCAI 2015, Volume 9351 of Lecture Notes in Computer Science. 234–241. Springer, Cham.

[258] Ira S. Rubinstein. 2013. Big data: The end of privacy or a new beginning? International Data Privacy Law. 3: 74.

[259] Waddah Saeed and Christian Omlin. 2023. Explainable AI (XAI): A systematic meta-survey of current challenges and future opportunities. Knowledge-Based Systems. 263: 110273.

[260] Jeffrey Saltz, Michael Skirpan, Casey Fiesler, Micha Gorelick, Tom Yeh, et al. 2019. Integrating ethics within machine learning courses. ACM Transactions on Computer Education. 19(4).

[261] Agariadne Dwinggo Samala, Xiaoming Zhai, Kumiko Aoki, Ljubisa Bojic, and Simona Zikic. 2024. An in-depth review of chatGPT's pros and cons for learning and teaching in education. International Journal of Interactive Mobile Technologies. 18(2): 96–117.

[262] Anke Samulowitz, Ida Gremyr, Erik Eriksson, and Gunnel Hensing. 2018. "Brave Men" and "Emotional women": A theory-guided literature review on gender bias in health care and gendered norms towards patients with chronic pain. Pain Research and Management. 2018(1): 6358624.

[263] Iqbal H. Sarker. 2024. LLM potentiality and awareness: a position paper from the perspective of trustworthy and responsible AI modeling. Discover Artificial Intelligence. 4(1): 40.

[264] Laura Sartori and Andrea Theodorou. 2022. A sociotechnical perspective for the future of AI: Narratives, inequalities, and human control. Ethics and Information Technology. 24(4).

[265] Peter Schaar. 2010. Privacy by design. Identity in the Information Society. 3(2): 267–274.

[266] Eva-Maria Schön, Michael Neumann, Christina Hofmann-Stölting, Ricardo Baeza-Yates, and Maria Rauschenberger. 2023. How are AI assistants changing higher education? Frontiers in Computer Science. 5.

[267] Reva Schwartz, Apostol Vassilev, Kristen Greene, Lori Perine, Andrew Burt, et al. 2022. Towards a standard for identifying and managing bias in Artificial Intelligence, Volume 3. US Department of Commerce, National Institute of Standards and Technology.

[268] Kate Seear. 2009. The etiquette of endometriosis: stigmatisation, menstrual concealment and the diagnostic delay. Social science and Medicine. 69(8): 1220–1227.

[269] Ramprasaath R. Selvaraju, Michael Cogswell, Abhishek Das, Ramakrishna Vedantam, Devi Parikh, et al. 2017. Grad-cam: Visual explanations from deep networks via gradient-based localization. *In*: 2017 IEEE International Conference on Computer Vision (ICCV). 618–626.

[270] Russ Shafer-Landau. 2018. Living Ethics: An Introduction with Readings, 3rd edition, Oxford University Press, 2024.

[271] Nima Shahbazi, Yin Lin, Abolfazl Asudeh, and H. V. Jagadish. 2023. Representation bias in data: A survey on identification and resolution techniques. ACM Computing Surveys. 55(13s).

[272] Kyarash Shahriari and Mana Shahriari. 2017. IEEE standard review—Ethically aligned design: A vision for prioritizing human well-being with Artificial Intelligence and autonomous systems. *In*: 2017 IEEE Canada International Humanitarian Technology Conference (IHTC). 197–201. IEEE.

[273] Jake Silberg and James Manyika. 2019. Notes from the AI frontier: Tackling bias in AI (and in humans). McKinsey Global Institute.

[274] Linda J. Skitka, Kathleen Mosier, and Mark D. Burdick. 2000. Accountability and automation bias. International Journal of Human-Computer Studies. 52(4): 701–717..

[275] Linda J. Skitka, Kathleen L. Mosier, and Mark Burdick. 1999. Does automation bias decision-making? International Journal of Human-Computer Studies. 51(5): 991–1006.

[276] Jessie J. Smith, Saleema Amershi, Solon Barocas, Hanna Wallach Jennifer Wortman Vaughan, et al. 2022. REAL ML: Recognizing, exploring, and articulating limitations of machine learning research. *In*: Proceedings of the 5th ACM Conference on Fairness, Accountability, and Transparency. ACM.

[277] Daniel J. Solove. 2002. Conceptualizing privacy. California Law Review. 90: 1087.

[278] Timo Speith. 2022. A review of taxonomies of explainable Artificial Intelligence (XAI) methods. *In*: Proceedings of the 2022 ACM Conference on Fairness, Accountability, and Transparency. 3531146. ACM.

[279] Ramay Srinivasan and Ajay Chander. 2021. Biases in AI systems. Communications of the ACM. 64(10): 62–71.

[280] Bernd Carsten Stahl and Mark Coeckelbergh. 2016. Ethics of healthcare robotics: Towards responsible research and innovation. Robotics and Autonomous Systems. 86: 152–161.

[281] Nisan Stiennon, Long Ouyang, Jeff Wu, Daniel M. Ziegler, Alec Radford Lehr, et al. 2020. Learning to summarize with human feedback. *In*: Advances in Neural Information Processing Systems.

[282] Julia Stoyanovich and Bill Howe. 2019. Nutritional labels for data and models. A Quarterly bulletin of the Computer Society of the IEEE Technical Committee on Data Engineering. 42(3).

[283] Emma Strubell, Ananya Ganesh, and Andrew McCallum. 2019. Energy and policy considerations for deep learning in NLP. http://arxiv.org/abs/1906.02243.

[284] Harini Suresh and John Guttag. 2021. A framework for understanding sources of harm throughout the machine learning life cycle. *In*: Proceedings of the 1st ACM Conference on Equity and Access in Algorithms, Mechanisms, and Optimization, EAAMO '21, New York, NY, USA. Association for Computing Machinery.

[285] Harini Suresh and John Guttag. 2021. A framework for understanding sources of harm throughout the machine learning life cycle. *In*: Equity and Access in Algorithms, Mechanisms, and Optimization (EAAMO '21). 1–9. New York, NY, USA. ACM.

[286] Latanya Sweeney. 2000. Simple demographics often identify people uniquely. Health (San Francisco). 671(2000): 1–34.

[287] Araz Taeihagh. 2021. Governance of Artificial Intelligence. Policy and Society. 40: 137–157.

[288] Telefónica S.A. 2018. AI principles of telefónica. Corporate Policy Document, Telefónica.

[289] The Guardian. 2023. Biased AI systems could be identifying images of 'racy' women's bodies, study finds. News article. Accessed: 2024-08-17.

[290] Judith Jarvis Thomson. 1976. Killing, letting die, and the trolley problem. The monist. 204–217.

[291] Judith Jarvis Thomson. 1985. The trolley problem. Yale Journal of Law and Technology. 94: 1395.

[292] Eric Trist. 1981. The evolution of socio-technical systems. Perspectives on Organizational Design and Behaviour. A detailed exploration of socio-technical systems' development from 1950 to 1970, including foundational concepts, early studies, and significant projects like the Norwegian Industrial Democracy project and the Shell Philosophy project.

[293] Laura D. Tyson and John Zysman. 2022. Automation, AI and Work. Daedalus. 151(2): 256–271.

[294] European Union. 2018. High-level expert group on Artificial Intelligence (AI-HLEG). `https://digital-strategy.ec.europa.eu/en/policies/expert-group-ai`.

[295] Scientific United Nations Educational and Cultural Organization (UNESCO). 2021. Recommendation on the Ethics of Artificial Intelligence.

[296] Bram Vaassen. 2022. AI, opacity, and personal autonomy. Philosophy and Technology. 35(4): 88.

[297] Heidi Vainio-Pekka, Mamia Ori-Otse Agbese, Marianna Jantunen, Ville Vakkuri, Tommi Mikkonen, et al. 2023. The role of explainable AI in the

research field of AI Ethics. ACM Transactions on Interactive Intelligent Systems. 13: 1–39.

[298] Ibo van de Poel. 2020. Embedding values in Artificial Intelligence (AI) systems. Minds and Machines. 30: 385–409.

[299] Aimee van Wynsberghe. 2021. Sustainable AI: AI for sustainability and the sustainability of AI. AI Ethics. 1: 213–218.

[300] Daniel Varona and Juan Luis Suárez. 2022. Discrimination, bias, fairness, and trustworthy AI. Applied Sciences. 12(12): 5826.

[301] Ashish Vaswani, Noam Shazeer, Niki Parmar, Jakob Uszkoreit, Llion Jones, et al. 2017. Attention is all you need. *In*: Advances in Neural Information Processing Systems, Volume 30. Curran Associates, Inc.

[302] Sahil Verma and Julia Rubin. 2018. Fairness definitions explained. *In*: Proceeding of the International Workshop on Software Fairness, FairWare '18. 1–7. New York, NY, USA. Association for Computing Machinery.

[303] Warren J. Von Eschenbach. 2021. Transparency and the black box problem: Why we do not trust AI. Philosophy and Technology. 34(4): 1607–1622.

[304] Ben Wagner. 2018. Ethics as an escape from regulation. from "ethics-washing" to ethics-shopping?

[305] Ethan Waisberg, Joshua Ong, Mouayad Masalkhi, Nasif Zaman, Prithul Sarker, et al. 2023. GPT-4 and medical image analysis: strengths, weaknesses and future directions. Journal of Medical Artificial Intelligence. 6.

[306] Wendell Wallach and Colin Allen. 2008. Moral Machines: Teaching Robots Right from Wrong. Oxford University Press.

[307] Hsiu-Ling Wang. 2024. Ability of bing chat to accurately answer pico clinical questions: a case study regarding preoperative oral carbohydrate intake. Taiwan Journal of Public Health. 43(1): 82–92.

[308] Xinpeng Wang, Xiaoyuan Yi, Han Jiang, Shanlin Zhou, Zhihua Wei, et al. 2023. ToViLaG: Your visual-language generative model is also an evildoer. *In*: Houda Bouamor, Juan Pino, and Kalika Bali, editors, Proceedings of the 2023 Conference on Empirical Methods in Natural Language Processing. 3508–3533. Singapore. Association for Computational Linguistics.

[309] Yutian Wang, Xuepeng Hu, Lingfang Yang, and Zhi Huang. 2023. Ethics dilemmas and autonomous vehicles: Ethics preference modeling and implementation of personal ethics setting for autonomous vehicles in dilemmas. IEEE Intelligent Transportation Systems Magazine. 15(2): 177–189.

[310] Anne L. Washington. 2019. How to argue with an algorithm: Lessons from the COMPAS-propublica debate. Colorado Technology Law Journal. 17(1): 131–160.

[311] Kellie Webster, Xuezhi Wang, Ian Tenney, Alex Beutel, Emily Pitler, et al. 2020. Measuring and reducing gendered correlations in pre-trained models. https://arxiv.org/abs/2010.06032.

[312] Laura Weidinger, Jonathan Uesato, Maribeth Rauh, Conor Griffin, Po-Sen Huang, et al. 2022. Taxonomy of risks posed by language models. *In*: 2022 ACM Conference on Fairness, Accountability, and Transparency, FAccT 2022, Seoul, Republic of Korea. 214–229. ACM.

[313] Chris Wiggins and Matthew L. Jones. 2023. How data happened: A history from the age of reason to the age of algorithms. WW Norton and Company.

[314] Aimee van Wynsberghe. 2015. Healthcare Robots: Ethics, Design and Implementation. Routledge, 1st edition.

[315] Yuxin Xiao, Shulammite Lim, Tom Joseph Pollard, and Marzyeh Ghassemi. 2023. In the name of fairness: Assessing the bias in clinical record de-identification. *In*: Proceedings of the 2023 ACM Conference on Fairness, Accountability, and Transparency, FAccT '23. 123–137. New York, NY, USA. Association for Computing Machinery.

[316] Xiaoyuan Yi, Jing Yao, Xiting Wang, and Xing Xie. 2024. Unpacking the ethical value alignment in big models. arXiv preprint arXiv:2310.17551. Under Review for Journal Publication.

[317] Liu Yu, Ludie Guo, Ping Kuang, and Fan Zhou. 2024. Biases mitigation and expressiveness preservation in language models: A comprehensive pipeline. *In*: Proceedings of the AAAI Conference on Artificial Intelligence. 38: 23701–23702.

[318] Muhammad Bilal Zafar, Isabel Valera, and Manuel Gomez Rodriguez. 2017. Fairness constraints: Mechanisms for fair classification. *In*: Proceeding of the Machine Learning Research. 54: 962–970. PMLR.

[319] Muhammad Bilal Zafar, Isabel Valera, Manuel Gomez Rodriguez, and Krishna P. Gummadi. 2017. Fairness beyond disparate treatment & disparate impact: Learning classification without disparate mistreatment. *In*: Proceedings of the 26th International Conference on World Wide Web, WWW '17. 1171–1180. Republic and Canton of Geneva, CHE. International World Wide Web Conferences Steering Committee.

[320] Rich Zemel, Yu Wu, Kevin Swersky, Toni Pitassi, and Cynthia Dwork. 2013. Learning fair representations. *In*: International Conference on Machine Learning. 325–333. PMLR.

[321] Xiaoming Zhai and Joseph Krajcik. 2022. Pseudo AI bias. arXiv preprint arXiv:2210.08141.

[322] Quan Zhang, Binqi Zeng, Chijin Zhou, Gwihwan Go, Heyuan Shi, et al. 2024. Human-imperceptible retrieval poisoning attacks in LLM-powered applications. 502–506.

[323] Wei Zhang, Xinyu Li, Hongyu Liu, and Xiaodong Wang. 2023. A brief survey on safety of large language models. CIT. Journal of Computing and Information Technology. 31(3): 207–223.

[324] Yunfeng Zhang, Q Vera Liao, and Rachel K.E. Bellamy. 2020. Effect of confidence and explanation on accuracy and trust calibration in AI-assisted decision making. *In*: Proceedings of the 2020 Conference on Fairness, Accountability, and Transparency. 295–305.

[325] Michael Zimmer. 2020. But the data is already public: on the ethics of research in Facebook. *In*: The Ethics of Information Technologies. Taylor & Francis.

[326] Erik Štrumbelj and Igor Kononenko. 2014. Explaining prediction models and individual predictions with feature contributions. Knowledge and Information Systems. 41(3): 647–665.

[327] Phillip Ball. 2014. Is AI leading to a reproducibility crisis in science? Nature. 624(7990): 22—25

[328] Info-communications Media Development Authority (IMDA) and Personal Data Protection Commission (PDPC). 2020. Model AI Governance Framework, 2nd Ed. Singapore.

[329] Arriagada-Bruneau, G., López, C. and Davidoff, A. 2025. A bias network approach (BNA) to encourage ethical reflection among AI developers. Science and Engineering Ethics. 31: 1. https://doi.org/10.1007/s11948-024-00526-9

[330] Lange, B., Keeling, G., McCroskery, A., et al. 2023. Engaging engineering teams through moral imagination: A bottom–up approach for responsible innovation and ethical culture change in technology companies. AI Ethics. https://doi.org/10.1007/s43681-023-00381-7

[331] Raja, A.K. and Zhou, J. 2023. AI Accountability: Approaches, Affecting Factors, and Challenges, in Computer. 56(4): 61–70. doi: 10.1109/MC.2023.3238390. keywords: Ethics; Artificial intelligence.

[332] Birhane, A., Steed, R., Ojewale, V., Vecchione, B. and Raji, I.D. 2024. AI auditing: The broken bus on the road to AI accountability. *In*: 2024 IEEE Conference on Secure and Trustworthy Machine Learning (SaTML). Toronto, ON, Canada. 612–643. doi: 10.1109/SaTML59370.2024.00037.

[333] Novelli, C., Taddeo, M. and Floridi, L. 2024. Accountability in artificial intelligence: what it is and how it works. AI and Soc. 39: 1871–1882. https://doi.org/10.1007/s00146-023-01635-y

[334] Goktas, P. 2024. Ethics, transparency, and explainability in generative ai decision-making systems: A comprehensive

bibliometric study. Journal of Decision Systems. 1—29. https://doi.org/10.1080/12460125.2024.2410042

[335] Buijsman, S. 2024. Transparency for AI systems: a value-based approach. Ethics Information and Technology. 26: 34. https://doi.org/10.1007/s10676-024-09770-w

[336] Gonen, H. and Goldberg, Y. 2019. Lipstick on a Pig: debiasing methods cover up systematic gender biases in word embeddings but do not remove them. *In*: Proceedings of the 2019 Conference of the North American Chapter of the Association for Computational Linguistics. 609—614. Minneapolis, Minnesota.

Index

Author Bios

Gabriela Arriagada-Bruneau is an Assistant Professor with double appointment in the Applied Ethics (IEA) and Mathematical and Computational Engineering (IMC) Institutes at Pontificia Universidad Católica de Chile. She is a philosopher in the same university, and holds a Master of Science in Philosophy from The University of Edinburgh. She is a doctor with a specialization in Applied Ethics, particularly ethics of AI and data at the University of Leeds. She is a Young Researcher at the National Center of Artificial Intelligence in Chile (CENIA) and is the Latin American Lead for the World Ethical Data Foundation (WEDF). She has participated in expert consultations to develop the first Chilean regulation of Artificial Intelligence and UNESCO recommendations on neuroethics. Her current research interest is in Bias and Fairness in AI, Feminist Philosophy of Science, and Ethics of Disabilities.

Claudia López works as an assistant professor at Universidad Técnica Federico Santa María in Valparaíso, Chile. Additionally, she is a principal researcher at the National Center of Artificial Intelligence (CENIA) and the Millennium Nucleus on the Futures of Artificial Intelligence Research (FAIR) in the same country. She earned her PhD in Information Science and Technology from the University of Pittsburgh. Claudia, who has a background as an Informatics Engineer, focuses her research on human-centred artificial intelligence, social computing, and human-computer interaction (HCI). Her current projects focus on understanding how citizens perceive AI, making the use of AI more transparent, and incorporating ethical considerations in AI development. Claudia actively promotes a socio-technical perspective of AI in public discussions and is deeply involved in initiatives to strengthen the research in HCI in Latin America and increase women's participation in technology development.

Marcelo Mendoza is an associate professor in the Computer Science Department at Pontificia Universidad Católica de Chile. He received a Ph.D. in Computer Science from the Universidad de Chile. He did a postdoc in Yahoo Research. He is a founder and former President of the Chilean Association of Pattern Recognition and President of the Chilean Foundation for Transparency and Democracy. He is a principal researcher at the National Center of Artificial Intelligence (CENIA) and associate researcher at the Millennium Institute for Foundational Research on Data (IMFD). His research is focused on developing AI methods to measure and predict events and processes involving humans. His work includes the study of information dissemination, the spatial distribution of humans within cities, community formation, and the analysis of constructive and destructive interactions among individuals. His contributions are significant in advancing the understanding of how AI can address complex societal challenges, particularly those related to the dynamics of human behavior. He has published over 100 peer-reviewed papers in conference proceedings and top-tier journals. In 2021, he received the Seoul Test of Time award for co-authoring the paper "Information Credibility on Twitter," a pioneering study in applying AI to the analysis of misinformation.